FAITH AND HEALTH

Endowed by

TOM WATSON BROWN

and

THE WATSON-BROWN FOUNDATION, INC.

FAITH AND HEALTH

Religion, Science, and Public Policy

Paul D. Simmons

MERCER UNIVERSITY PRESS

MACON, GEORGIA

MUP/P369

1400 Coleman Avenue
Macon, Georgia 31207

First Edition.

Books published by Mercer University Press are printed on acid free paper that meets the requirements of American National Standard for Information Sciences—Permanence of Paper for Printed Library Materials.

Library of Congress Cataloging-in-Publication Data
Simmons, Paul D.
Faith and health : religion, science, and public policy / Paul D. Simmons. —1st ed.
p. cm.
Includes bibliographical references and index.
ISBN-13: 978-0-88146-085-8 (pbk. : alk. paper)
ISBN-10: 0-88146-085-0 (pbk. : alk. paper)
1. Suffering—Religious aspects—Baptists. 2. Health—Religious aspects—Baptists. 3. Medical care—Religious aspects—Baptists. 4. Medical ethics—Decision making. 5. Christian ethics—Baptist authors. 6. Religion and science. I. Title.
BT732.7.S58 2008
201'.7621--dc22

2007051524

The Painting of Balbases: Sts. Cosmas and Damien

Oil painting on panel 168 x 133 cm. attributed to the Master of Los Balbases, Burgos, Spain, c. 1495 Iconography. The panel depicts the amputation of an ulcerated leg of a Christian verger, and the attachment of a non-diseased leg from a dead Moor. The painting is sometimes attributed to Alonso de Sedano, a contemporary of Balbases. Another panel of the six in the altar piece grouping depicts the martyrdom of Cosmas and Damian. They were beheaded for the blasphemous act of trying to correct an illness thought to have been sent by God on the man as punishment for sin. The Wellcome Library, London. Used with permission.

In Memoriam

Henlee Hulix Barnette (1912–2005)

Teacher and professor extraordinaire, mentor, friend, and colleague, who thought it more important to be prophetic than to remain safely conservative; more important to seek truth than to settle for comfortable platitudes; and more important to be inclusive toward the different and despised than to join the ranks of the powerful who exploit the vulnerable and make bigotry an article of faith.

Dedicated to

Brent and Connie, Brian and Catherine, three children and a daughter-in-law whose questions and persistence about life in the critical stages present the challenges and issues that this book is all about.

CONTENTS

LIST OF ABBREVIATIONS

AI: Artificial Intelligence

ALS: Amalytropic Lateral Sclerosis (Lou Gehrig's disease)

AMA: American Medical Association

BABI: Blastomere Analysis Before Implantation

COPD: Chronic Obstructive Pulmonary Disease

CPA: Certified Public Accountant

CPR: Cardio-pulmonary Resuscitation

CTA: Composite Tissue Allotransplantation

DNR: Do Not Resuscitate

EBV: Epstein-Barr Virus

EMTALA: Emergency Medical Treatment and Active Labor Act

ESRD: End-stage Renal Disease

GDP: Gross Domestic Product

GNR: Genetics, Nanotechnology, and Robotics

HMO: Health Maintenance Organization

HCR: Hastings Center Report

IFSSH: International Federation of Societies for Surgery of the Hand

IOM: Institute of Medicine

IRB: Institutional Review Board

IVF: In Vitro Fertilization

JAMA: Journal of the American Medical Association

JCS: Journal of Church and State

JLME: Journal of Law, Medicine, and Ethics

MCO: Managed Care Organization

NEJM: New England Journal of Medicine

PAS: Physician-assisted Suicide

PGD: Pre-implantation Genetic Diagnosis

PSDA: Patient Self-determination Act

PTLD: Post-transplant Lymphoproliferative Disorders

PVS: Persistent Vegetative State

TAH: Total Artificial Heart

TIAH: Totally Implantable Artificial Heart

VAD: Ventricular Assist Device

INTRODUCTION

This book has been written with several objectives in mind. First, it is a response to contemporary debates about bioethical issues. Eleven essays address a variety of topics, from the purposes of medicine to issues posed by technological developments. Technology not only changes the options regarding medical treatments, but poses challenges for ethics and religious understandings of what it means to be human.

Another purpose of the book is to take seriously the challenge of politicized evangelicals regarding the relation of religious faith to public policy. They are correct in saying that faith should guide decisions not only personally but in the way we influence and fashion public policy. But how should that be done in a pluralistic society? Many of us believe that those most determined to "change America" are leading in the wrong direction and for many of the wrong reasons. All Americans have a place at the table and should share in both the benefits and burdens that go with being a citizen. As a nation, we are governed by the Constitution, not by religious groups or doctrinal opinions.

The Scriptures are dear to the devotional life and a guide to both personal and public behavior for religious people seeking to redeem the social order, but public policy is never simply a matter of reading from the Bible and deciding what legislative or personal responses should be made to particular issues. Those who believe in biblical authority have strong disagreements among themselves about what is taught, much less how its teaching should be implemented in a society that is very different from the one in which the injunctions were originated.

A third purpose of the book is to give guidance regarding issues of method in decision-making. How people go about making decisions is a subject of considerable importance. Everyone brings certain assumptions and beliefs to the decisions they make, but few take the time to analyze the process at work as they engage the issues. This book attempts to draw attention to the major variables in ethical decision-making. One of the crucial elements will be beliefs about the proper relationship between science and faith. How does one reconcile seeming differences in perspective between the sacred texts of religion and the findings of science?

Medicine and Faith: Cosmas and Damien

The portrait on the cover captures the dynamic interplay between religion, science, and public policy, the basic concern of this book. The work is now attributed to the Master of Los Balbases, by the prestigious Wellcome Library, London. Others have attributed it to Alonso de Sedano, whose works are also found among the panels in the Cathedral of Burgos, Spain. Dated circa 1495, the panel depicts the amputation of a cancerous leg of a Christian verger and the attachment of a non-diseased leg from a dead Moor. Other panels of the six in the altar group depict the martyrdom of Cosmas and Damian, the physicians in the painting who became the patron saints of surgery. They were beheaded for the blasphemous act of trying to correct an illness. Orthodox doctrine taught that disease was sent by God as punishment for sin.

The painting has it all: medical science, religion, and public policy. The picture is a profound but disturbing reminder of the acrimony and injury that often attends confrontations between religion and science. The surgeons are the center of attention in the painting since they resisted the church's efforts to prohibit the explorations of medicine. They were heroes in the annals of the courageous who resisted the arrogance of power so often manifested by religious leaders. The physicians were not anti-religious, but instead people of faith as they extended the boundaries of medical knowledge. Ironically, they became "saints," as the halo and the presence of angels indicate. The blessings of God were on the transplant, according to the artist. All the features of the contemporary debates about religion, medicine, and ethics come together in this fifteenth-century portrait.

The question persists, of course, as to how physicians should decide whether to use a medical treatment that is innovative and controversial. The Balbases portrait captures the human desire for wholeness after the assault on bodily integrity and health by a dreadful disease. Religious leaders raised issues of faith and piety and objected to medical interventions based on beliefs that certain actions in medicine would be impious and blasphemous. Should the doctor accede to the objections of clergy or follow their own impulse to treat disease? Might the sense of religious truth and moral values on the part of the surgeon and the patient prevail, or should they defer to the officials of the church? The fact that they could be excommunicated makes the question even more complicated and critical. The threat of excommunication is extremely coercive for those who take the church and their salvation seriously. That point was not lost on those who noted the threats by some prelates to excommunicate any politician who supported reproductive choice and elective death contrary to official church teachings.

Very likely, both clergy and physicians shared the patient's hope for healing. The clergy thought in terms of miracle and a piety that merited divine intervention. The physician hoped for medical knowledge and transplantation skills. The patient's concern was for healing. He was likely indifferent as to how it happened but deeply invested in wanting it to happen. His pain and suffering cried out for relief, but he was caught in the trap of fearing religious rejection while seeking medical assistance.

Science has often had to follow different paths than those approved by the church. Religious leaders defend certain tenets of faith as if they were defending the truth of God. Attacking and resisting science may be regarded as the demands of piety and fidelity to God. Men and women of science have suffered from the collusion of church with government to ban procedures thought to be in open opposition to the teaching of orthodox doctrine or made controversial by religio-political action groups. All patients now benefit from those who suffered persecution by the church in order to advance medical science. Physicians followed a different vision because they saw the issues in very different ways than clergy. The world is deeply indebted to those willing to pay the price of religious condemnation and ecclesiastical outrage. Fortunately, some physicians were equipped with the courage necessary to follow the medical mandate in the interests of human health and well-being.

Some forms of religion still attempt to dominate science and dictate public policy. Their tactics aim to control and hinder innovative medicine. Galileo and Copernicus suffered at the hands of the medieval church, which thought the faith of believers would be undermined if the discovery that the world actually circled the sun (when the opposite was previously accepted as truth) gained wide acceptance. The church used pre-Copernican astronomy as a way to enforce its teaching that God had created the world for the benefit of people. The end result was that people were the only value—all other creatures were expendable, and nature was subject to endless exploitation.

The temptation to do battle with the proposals or possibilities of science has a powerful appeal on the current scene. We need only be reminded of the battle over experimenting with embryo stem cells (see ch. 9). Scientists believe such cells hold the promise of major breakthroughs in treating certain forms of illness and the effects of devastating injury, but groups on the religious right, claiming to have special insight from God, oppose the scientific investigations that might make such therapeutic interventions possible.

The religious argument against stem cell research is that it would be against the will of God for such scientific endeavors to be permitted. President Bush made this point explicitly in his statement banning the use of federal funds for such research. For him, it was a religious and moral matter. Religious groups brought no medical or scientific evidence to the debate. Rather, they sought to control the issue with moral and political influence based on a certain view of orthodox doctrine. (See Appendix I, which presents a graphic outline of various sources of authority in medical decision-making.) The Pontiff, along with ultraconservative evangelicals, insisted that public policy was to embody and protect religious doctrine and thus limits had to be placed on medical research. The debate continues.

Terri Schiavo: Religion, Medicine, and Politics

The tragic case of Florida resident Terri Schiavo also captures the concerns of this book. Schiavo's story embodies issues in medical ethics and the ways in which religious beliefs, moral values, political action groups, and the courts affect decisions about medical interventions at the end of life.

The Schiavo case became a contest over whose religious values and sense of truth should prevail—those of the patient and physicians committed to her care, or those who claimed special religious and moral insight as to actions that should be prohibited. Religiously motivated political action groups forced an extremely private and personal matter into the political arena. The disagreements were between people on two sides of an issue, and they all claimed religious values as grounding for their decisions and actions. The outcome was a public display of power that seemed to care little for Schiavo as a person. They disregarded her faith commitments and wanted to impose a religious and moral opinion on her. It was also an attack on the courts since her husband was carrying out her wishes regarding aggressive medical treatments when she could no longer speak for herself.[1]

Schiavo had been in persistent vegetative state (PVS) since her heart failed in 1990 at age twenty-six, apparently from a chemical imbalance related to anorexia. (See Appendix VII: "Timeline in the Terri Schiavo Case.") She was resuscitated, but the absence of oxygen from the brain

[1] The court had settled the question as to whether a person could refuse treatment in the *Cruzan* case in 1989.

caused extensive damage. She had permanently lost cognitive or decisional capacity. Her husband sought medical treatments from coast to coast, but her condition was irreversible.

Three years after her lapse into a persistent vegetative state, conflict over who should be making decisions as her proxy resulted in a lawsuit by her parents, Bob and Mary Schindler. They wanted her husband, Michael, removed as the guardian. The court threw out the case. A spouse of an incompetent partner is the guardian under Florida law, unless it can be shown that there has been abuse, neglect, or malfeasance. The guardian *ad litem* found in Michael's favor, as did the court.

Michael appealed to the court for permission to withdraw the gastronomic tube and allow Terri to die a natural death ten years after her lapse into a persistent vegetative state, but the Schindlers continued their opposition and insisted that Terri be aggressively treated in spite of her extensive impairment.

Jay Sekulow, attorney for the American Center for Law and Justice (ACLJ), persuaded the Schindlers to make Terri's situation a test case on the legal right to refuse treatment.[2] The case went through every level of the judicial system in Florida and was finally appealed to the US Supreme Court, which refused to hear it. A federal judge allowed the tube to be reinserted, however. Terri was reintubated, only to have the tube removed a second time.

Politicians then entered the case to challenge the court's order. Florida governor Jeb Bush signed a hastily prepared bill and ordered the tube to be reinserted. The Florida Supreme Court then struck down "Terri's Law" as unconstitutional, and permission was again granted to disconnect the feeding lines. The US Congress also became involved on the Schindlers' side. They called it being "on the side of 'life,'" which meant they wanted to curry the favor of the religious right.

Meanwhile, Terri's tube was removed for the third time, causing quite a frenzy on Capitol Hill. President Bush flew back from Texas to sign the bill for the Relief of the Parents of Theresa Marie Schiavo Act[3]. Senators flew in from their vacations around the world. The Schindlers filed an emergency request with a federal judge to have the tube reconnected, but

[2] The ACLJ is an arm of the Christian Coalition, founded by Pat Robertson, and often takes positions opposite those of the ACLU (American Civil Liberties Union), which was by this time representing Michael.

[3] S.686 ES.

the 11th Circuit Court of Appeals in Atlanta refused. Terri died thirteen days after the tube was removed.

Terri's story was covered extensively by the media, and the country became deeply divided over the issue of whether or not she should be sustained with aggressive medical treatments. The autopsy showed that her brain had deteriorated to half its normal size. In short, she had lost all mental capacity that would have enabled her to respond to, or even be conscious of, people around her. Emotional appeals to the contrary were divorced from all medical evidence as to her neurological or mental capacities.

The evidence considered by all the courts indicated the good-faith efforts of her husband, Michael. There was no reasonable doubt that she would, in fact, have wanted the tubes removed.[4]

The personal and political acrimony that led to extensive hearings and repeated intubations were injurious to Terri. She undoubtedly experienced brain damage in addition to her initial insult from the multiple withdrawals of food and nutrition. Ugly and untrue allegations were made against Michael, and enormous expense was involved by the legislative efforts (both state and federal) to overturn the courts' decision.

The interventions by politicians and conservative religious groups seemed little more than an egregious abuse of power. The battle between the Schindlers and Michael escalated to confrontations between the executive and legislative branches of government against the judicial, at both state and federal levels.

Even the ill Pope John Paul II spoke out against the withdrawal of Terri's gastronomic tubes.[5] Ironically, he was also dying but refused to return to the hospital, choosing to remain in his own quarters and die without life-sustaining interventions. He wanted Terri Schiavo to be maintained, but he did not want to have such procedures for himself.

The Schiavo case now has significance for medical ethics that goes beyond the simple facts and details of suits and countersuits. It is a classic case of confrontation among people motivated by religious or faith commitments who see the issues of life and death in very different ways. One side insists on rights to bodily integrity and individual liberties; this side

---

[4] See Lawrence O. Gostin, "Ethics, the Constitution, and the Dying Process: The Case of Theresa Marie Schiavo," *JAMA* 293/19 (May 18, 2005): 2403.

[5] Address by Pope John Paul II to the International Congress on "Life-sustaining Treatments and Vegetative State: Scientific Advances and Ethical Dilemmas," March 13, 2004. (Available online.)

supported Terri and her right to say no to life support machines, and they supported Michael's efforts to implement Terri's wishes. The opposing voice argued strongly that any decision that invites or cooperates with death is evil and should not be allowed. They further insisted on their right to impose their views on everyone through public policy.

A Question of Method in Ethics

This book examines the right to die (as in the Schiavo case) and other issues in ethics and medicine. Ethics analyzes the arguments and the interplay of various assumptions behind a given defended position or pursued objective. That is the question of method in decision-making.

Method is the manner in which one relates the various components or variables that comprise a particular decision. Everyone has a way of making decisions. There is an inner consistency or cohesion of thought that results in the conclusion one reaches. Few people give much thought to the process, but they nonetheless relate internal beliefs, motives, intentions, the "facts" of a case, and religious perspectives in a way that can be analyzed and understood.

Such reflective dialogue is required in a society that values democratic processes and public policies that serve the interests of all citizens. Knowing what a heavy-handed and coercive movement is up to may lead to concerted action to resist their goals and change directions for the society they are trying to dominate. Authoritarian religion seeks to deprive others of power over their own lives or equal rights to act on personal beliefs or moral principles. The Nazi demonism appealed to ideological perspectives that denied basic human rights. Resisting the Nazis was morally required. Tragically, many people either did not understand or did not care about the extremism at work.

Four factors in method. Every debate in ethics involves the interplay of four major factors: perceptions of the facts or data, moral rules or principles, faith loyalties, and ground-of-meaning beliefs.[6] One's perception of the data takes account of the medical "facts" and how they are related to the judgments being made. Facts or data are seldom simply objectively considered or understood. Few people are able perfectly to transcend subjective elements in decision-making. For instance, two neurosurgeons may disagree about the most beneficial medical treatments for a patient

---

[6] Glenn H. Stassen, "Critical Variables in Christian Social Ethics," in *Issues in Christian Ethics*, ed. Paul D. Simmons (Nashville: Broadman Press, 1980) 57.

while viewing the same X-rays or other diagnostic studies. Both are specialists, highly trained people of intelligence and good faith. The disagreements likely reflect different *perceptions* of the data.

The use of moral rules and principles is another persistent factor in medical ethics. The emphasis on rules that are absolute and non-negotiable has been very much in evidence in conservative religious circles that now dominate the debate in America. At this level, rules are given to answer the moral question about the obligations owed or the actions forbidden. No thought is given to whether the rule fits or that the assumptions involved stretch logic to the point of incredulity. In the stem cell debate, for instance, the rule most often given against experimentation is that of "do not kill human life." The critical question, of course, is whether a rule that protects people is also to be applied to stem cells, and, if so, for what reasons and in what way.

The principle approach avoids the rigidity and absolutism of rules by emphasizing general statements of moral obligation. Moral obligation is focused by thinking through the requirements of the principle at stake.

Principlism in Medical Ethics

Principlism is arguably the most widely used approach to medical ethics in America. Four principles constitute the points to be considered in analyzing the moral obligation associated with particular cases in medicine.[7] Autonomy, or respect for the person, relates to the patient's prerogatives to govern one's own life in terms of one's own internalized values and religious beliefs. In that sense, it is individualistic. The central moral value is that persons and their religious perspectives are to be respected and protected. Persons are self-defining moral agents who have the capacity for thinking through their own dilemmas and deciding a course of action based on values and beliefs that define the self. Patient preferences[8] thus govern decisions in a range of issues, from reproductive rights to end-of-life decisions.

The second principle is beneficence, which relates the goals of medicine to the health needs of the patient. The good of the patient, not the self-interests of the physician or healthcare institution, is of utmost

---

[7] See T. L. Beauchamp and J. L. Childress, *Principles of Biomedical Ethics*, 5th ed. (New York: Oxford University Press, 2001) 12.

[8] See Albert Jonsen, Mark Siegler, and William Winslade, *Clinical Ethics*, 6th ed. (New York: McGraw-Hill, 2006), who prefer to approach ethics in the clinical setting around the topics of medical indicators, patient preferences, quality-of-life considerations, and contextual factors.

importance. Beneficence aims to treat the patient based on medical need, not moral merit. Even patients who have an addiction to smoking or alcohol that compromises their health deserve competent, compassionate medical treatment.

The third principle is non-maleficence, or the obligation to "do no harm, is the "other side" of beneficence. The Hippocratic Oath emphasizes the first rule of medicine, which is to "do no harm." Enthusiasm to treat may overlook evidence that a particular intervention is non-beneficial, thus running the risk of injury to the patient. Caution in medical experimentation is also related to the commitment to guard the well-being of the patient. The principle of non-maleficence was given great prominence in the trials of Nazi physicians. Ideology, not beneficence, prevailed in the experimentations carried out against Jews, gypsies, the mentally ill, criminals, and others. Medicine distorts its moral mandate and loses scientific integrity when it serves the interests of ideology instead of the person.

Justice, the fourth principle of medical ethics, seeks a fair distribution of the benefits and the burdens of medicine. All sides in a debate will hold some concept of justice, but the meanings vary widely. Karen Lebacqz has outlined six theories in current debates about social issues from health care to poverty.[9] Each theory has a central contention, but a variety of perspectives exists within the general approach.

Three theories of justice get special attention in American bioethics. Utilitarianism seeks to maximize a central value, either for the self or for the greatest common good. The question is always how to balance social costs with personal benefits. In other words, how might we make social arrangements that enhance each person's right to "life, liberty, and the pursuit of happiness"?[10]

Egalitarianism stresses the equal value of each person and thus establishes a claim for equal treatment where questions of distribution or enjoyment of benefits are concerned. Human rights language appeals to entitlements belonging to every person. In religious terms, each person is regarded as special to God in creation and redemption and thus has something like an equal claim to social benefits.

Libertarianism stresses individual rights to property and to personal freedoms that should not be limited by government or other powerful institutions, such as the church.

[9] Karen Lebacqz, *Six Theories of Justice* (Minneapolis: Augsburg Press, 1986).
[10] United States Declaration of Independence, adopted 4 July 1776.

Needless to say, disagreements about the superior approach persist among persons equally committed to noble goals and what they regard as human rights or religious truths. The very fact of continued disagreements shows the failure of rational principles to achieve general agreement among thoughtful people. Principles are necessary but hardly sufficient to settle the thorny issues in medical ethics. Like rules, principles also have a variety of sources and thus generate opposition from those who feel their interests are being violated or ignored.

Other Approaches to Medical Ethics

Some object to the emphasis on principles, believing they emphasize reason to the neglect of more powerful factors or forces in decision-making, such as emotion, a special point of view, or a particular religious story. Narrative ethics emphasizes the particularity of community beliefs or of a personal history.

The women's movement emphasizes feminist ethics or an ethics of care, believing the history of Western thought has been dominated by men who do not understand the uniqueness of women's perspectives in moral and medical matters. Feelings, sympathy, compassion, love, and friendship may be more compelling than any moral or philosophical argument. The woman's perspective is especially important in reproductive issues since she has unique capacities and bears an unequal burden in childbearing.

Virtue ethics emphasizes traits of character that establish internal moral boundaries that the person will not violate. A truthful person will not tell lies, nor will a loving person commit acts of violence. Certainly it is desirable that physicians be persons of upright character and integrity. Virtue ethics has an instant appeal since it expresses the ideal person or character everyone could trust and who would never knowingly do harm.

But an emphasis on virtues is hardly sufficient in the world of medicine. A major problem is that virtues begin to function as rules and their proponents as authoritarian leaders. Further, the stresses are too extensive, the temptations too subtle, and the medical diagnoses too fallible to trust only in the person's own capacity for knowing and doing what is right and true. Principles and rules—legal, moral, and professional—are necessary to remind us all of the integrity of thought and action required of those who are entrusted with the care of people in need of medical assistance. Professional standards not only reflect an important consensus as to reasonable expectations, but help to define the virtues necessary and admired within any profession.

Casuistry, or case-based approaches to ethical dilemmas, attempts to integrate the insights of law, medicine, and psychology.[11] The "case" in one area has continuity with a case in other areas, and each can learn from the other. The principles of interpretation that are employed are derived from the case itself, or from the wisdom gleaned from a history of similar cases, and they establish a general moral framework for understanding. Historical precedent may be more important than abstract principles intended to express a moral obligation.

Faith: Loyalties and Interests

Faith and its loyalties go beyond social benefits such as harmony and peaceful co-existence. People have very different basic beliefs—about life and death, about the meaning of life itself, and about the basic reason for being moral. Tolerating differences among people of religious faith seems a worthy goal of public discourse and a necessary procedural norm for political and social harmony. Those who believe they have a divine calling to change the world into their image of goodness and virtue, however, typically regard tolerance as unacceptable compromise. Some things require absolute loyalty, according to their approach to religion and ethics.

American evangelicalism, for instance, began with a declaration to "do battle" to preserve the "fundamentals" of the faith.[12] The special enemies cited were evolutionists and biblical critics. They were targeted as special objects of resistance since they were invading mainline seminaries and universities.

Thus, philosophical or theological commitments enter the debate as answers to fundamental questions. James Dickey once said the most important question one will ever ask is, "What is your life all about?" The fact that people answer the question in different ways explains a great deal about ethical disagreements and the relation of religion to politics. Those most aggressive and belligerent toward others may couch their actions as inspired of God and thus hardly open to question. They can be dangerous people.

---

[11] Albert Jonsen and S. Toulmin, *The Abuse of Casuistry: A History of Moral Reasoning* (Berkeley: University of California Press, 1988).

[12] Gabriel Almond, R. S. Appleby, and E. Sivan, *Strong Religion: The Rise of Fundamentalisms around the World* (Chicago: University of Chicago, 2003) 2, citing an editorial in *The Watchman Examiner*, 1920.

The Bible and the Approach of This Book

The method embraced in this book brings together the principles of medical ethics and religious beliefs, including insights from biblical teachings. Using the Bible in medical ethics can be extremely problematic, of course.[13] The Bible is neither a book of rules for human behavior nor a definitive dictionary of medical ethics. It can be used in a way that alienates those who question its validity as a guide for moral behavior or have other sources as sacred scripture. I do not appeal to Scripture to settle the issue in moral debates among Christians, but as a way to integrate biblical perspectives in moral and theological reasoning. Biblical perspectives, stories, and aphorisms are commonplace in the discourse of Americans—even among people who claim no particular faith loyalty to Abrahamic religions.

Just how the Bible is used or how it may be misrepresented in various arguments is a major concern of this book. Those in the "religious right" present challenges as to the meanings of faith, the content of biblical teachings, and the methods and goals of social change. They are correct to believe that faith is important in the way believers intend and attempt to influence the society in which we live, but are they correct on the issues they embrace as central to Christian faith? Are they correct in the manner they attempt to bring about social change?

A major component of discussions about medical ethics is a debate among Christians, most of whom accept biblical authority but disagree strongly about what it teaches and what faithfulness to biblical perspectives actually requires or prohibits. Dealing with ethics in America requires attention to the Bible, in part at least because religious groups appeal to the Bible to justify egregious actions and employ actions that seem both undemocratic and non-biblical.

Aggressive religious leaders do not like criticism, of course, and often insist that their point of view is above dispute or that they represent the (only) *Christian* perspective. Whether such claims are true or arrogant and self-serving is necessary to explore. Simply because a movement has large numbers of prominent religious leaders is no sign that it is to be trusted or followed. After all, we have numerous cases of misguided religious leaders who have betrayed central tenets of the Christian faith and led their followers to their death or to doing damage to society as a whole. Jim Jones

---

[13] See A. Verhey, *Reading the Bible in the Strange World of Medicine* (Grand Rapids, MI: Eerdmans Publishing Co., 2003) and C. H. Cosgrove, *Appealing to Scripture in Moral Debate* (Grand Rapids, MI: Eerdmans Publishing, Co., 2002).

and David Koresh are strong reminders of the fact that religion can go bad. Both leaders appealed to the Bible and claimed to be motivated by a Christian calling, but there were good reasons to think they were seriously misguided.

My use of the Bible is as a primary source of guidance or wisdom in terms of the human involvement in social and scientific issues that are important for each of us. Using the Bible is not to claim an absolute authority for any perspective that might be embraced or advocated. The Bible is neither a book of rules nor moral laws that are easily discernible and obviously applicable to contemporary issues. But the moral imagination finds a rich reservoir of images, stories, and metaphors in Scripture that enrich and give direction to the search for meaning and obligation in human relationships.

The Bible has a place of special importance as sacred literature in the Christian, Jewish, and Muslim traditions, of course, but all people are welcome to ponder its insights and learn from its stories, aphorisms, and moral appeals, regardless of their views of its inspiration and authority. My effort has been to use Scripture to illuminate the human situation and provide wisdom for guidance as people seek their way through the ethical conundrums at the heart of medicine. The Bible is not a collection of prescriptions for curing all the world's ills. One cannot turn to the Bible and find clear and unambiguous answers to the great ethical quandaries from health care to abortion. Even so, it is indispensable to Christians and many others as a source of guidance for ethical reflection.

The religions of the world have a great deal in common and have a great deal to learn from one another. Problems among and between religious groups grow out of misunderstandings and rigid claims of superiority/inferiority by one toward another. Religious intolerance has the malevolent face of bigotry, not the gentle persuasion of truth, nor the persuasive embrace of compassion. The fact that the Bible has been used to oppress minorities and justify injurious behavior from ethnic cleansing to child abuse is a reminder that ethical commitments are brought to, not simply gleaned from, the Scripture.

Religious Liberty: The Social Contract

Another important component in religious and medical ethics is the crucial role of human rights. Religious liberty is rightly regarded as the first freedom and a fundamental human right. The separation of church and state is a corollary to religious liberty and fashions the political posture necessary to guard against coercion in religious matters through public policy. A study of the history of the church and its relation to the state in various eras shows that enormous excesses have emerged over the centuries when the interests of institutional religion and those of government become entangled, confused, or identified with one another. The histories of the Holy Roman Empire, the Inquisition, the era of "Conformity" in England, and the Puritan experiment in America are continuous reminders of the bad things that happen when religion and government get too closely aligned or entangled.

The Baptist story in which I have been nurtured is one that champions the vision upheld by Roger Williams, first governor of Rhode Island. He also established the first Baptist church of Providence. Governance in that colony was committed to protecting liberties of conscience for all citizens. No government coercion was allowed in religious matters in any way—whether pressure to attend church, believe certain doctrines, or regulate practices related to family life. Women were protected equally with men as citizens. Atheists and agnostics were free not to be religious. Catholics, Muslims, Jews, and Deists were free to believe according to their own persuasion and personal conscience. Tax monies were not used for religious purposes. Government stayed out of the business of religion.

That pattern was adopted by the Constitutional Congress. Under the influence and leadership of Thomas Jefferson, James Madison, Benjamin Franklin, and others, protections for all Americans related to religious liberties were instituted. The First Amendment in the Bill of Rights declares, "Congress shall make no law respecting an establishment of religion nor prohibiting the free exercise thereof." Americans are still trying to discover and live out the full implications of that powerful statement of fundamental human rights, but religious liberty—and its corollary, the separation of church and state—is one ideal that makes America the land of freedom.

Religion and Science.

My commitment to religious liberty figures prominently in my concerns about "faith" and "public policy." It also figures into the "science" component of this book since conflicts between religion and science have generated heated debate and created problems historically for both faith and science, as in the Galileo episode. On the current scene, arguments about public policy and embryo stem cell research have posed the issue strongly as have the debates over abortion, the right to die, and physician-assisted suicide.

The problem of faith and public policy is also at issue in questions pertaining to Intelligent Design and other forms of Creation Science. The anti-science attitude of fundamentalism and the authoritarian approaches of various religious traditions continue to generate problems for educators and scientists, most of whom can tell the difference between religious beliefs and scientific findings. The critical issue is how to relate religious perspectives or doctrines to the data of science, and how to relate science to the study of sacred texts.

The Place and Importance of Experience

To a considerable extent, I belong to that group of ethicists described as "engaged in civil rights and anti-war activities" early in their professional careers and "trained in Protestant theology and religious ethics (who) carried their...individualism, anti-paternalism, and anti-authority convictions into bioethics."[14] The features outlined by René Fox and others cover a wide range of thinkers and writers in medical ethics but give little help when finding differences among us. We take divergent stands on many issues for some of the same reasons cited as explaining our involvement in bioethics.

Various strands of thought and influence will be seen in the essays found in this book. My personal experiences figure prominently in the topics selected, the cases explored, and the positions argued. Some cases are drawn from newspapers or magazines and thus belong to the public domain. Cases drawn from the hospital setting have been altered sufficiently to protect privacy but described in sufficient detail so as to retain the integrity of the points to be made. In all these areas, my work reflects my theological, philosophical, and pastoral interests.

A physician friend in a town where I served as pastor helped shape my interest in medical ethics. Long before medical ethics became a growth industry, we were discussing the issues: women needing abortions when the law was terribly restrictive; the poor (especially African Americans) in town who were arbitrarily excluded from civic and health benefits by having city limits drawn so as to deprive them of water and sewage services; segregated

---

[14] C. M. Messikomer, R. C. Fox, and J. P. Swazey, "The Presence and Influence of Religion in American Bioethics," *Perspectives in Biology and Medicine* 44/4 (Autumn 2001): 492.

facilities with signs in the windows of businesses that said "whites only"; families who could not afford the fees for medical services; and families devastated by the sudden death of a child in an auto accident or grief-stricken by the onset of a lethal and incurable illness.

The manner in which people die often raises ethical issues. Usually, people in "our town" died from advanced age and incurable illnesses. Some deaths were sudden, some slow and painful, but almost always with the counsel of both minister and physician. His own tragic life and death were reminders of the inadequacy of our resolutions about healthy living and responsible lifestyles.

The faces and stories of people I have known as pastor, professor, friend, or acquaintance figure prominently for me in the way issues in ethics are framed and resolved. Their faces convey the anguish and frustration of coping with difficult issues, especially when badgered by people who do not share their pain but have easy answers and quick judgments. Some are children impregnated by relatives or students violated and impregnated by rapists or molested by faculty or other trusted persons. Others show the anxiety and fear associated with dying an ugly and painful death. Other families faced the psychic pain and constant burden of caring for children with congenital defects or incurable disease. Their stories are written in anguish and hope, and they can be multiplied by the dozens.

Out of such experiences, I have learned the inadequacy of facile and absolute answers in the midst of complexity and ambiguity. I also have learned the indispensability of biblical wisdom for the comfort and guidance it provided to those familiar with its pages and open to its counsel. I recoiled at stories of pastors who had easy answers dispensed with glib responses to stories of human tragedy and encounters with the great issues of life. The challenge is always to study the Scriptures so as to "rightly discern" its wisdom and truth.

The Organization of the Book

This book begins and ends with chapters that deal with the problem of evil. The opening chapter explores the nature and purpose of medicine around the topic "Human Suffering and the Medical Mandate." The last chapter explores the question of demonic possession and the prospect of exorcism as a medical intervention. Between those chapters are explorations of various problems or issues that pose the problem of human suffering in different but related ways.

Chapter 2 deals with "False Hope: Faith & Healing in the ICU," exploring the problem of discerning the difference between hopes that facilitate health and healing and those that deny reality and exact a heavy toll in medical expenses and human energies trying to salvage those doomed by their physiological and/or neurological burdens.

Chapter 3 deals with "Health Care and the Future," reflecting on the prospects for a more inclusive healthcare system that will more perfectly embody concerns for justice and equality. We are faced with a system that does not deliver what it promises but is more expensive than any other in the world. Reform is needed, and now.

Chapter 4 explores the problem of "Aging as an Assault on Human Dignity: Spirituality and End-of-life Decision-making," starting with a struggle with the meaning of human dignity. I suggest a phenomenological approach, which avoids the positivism of simply affirming dignity without defining or explaining its presence or absence in human life.

Chapters 5 and 6 continue the discussion of end-of-life issues, first exploring the relation between "Religion, Politics, and a Right to Die" followed by "Faith and the Future of Physician-Assisted Suicide." Questions about just what a right to die means or involves and how such a right might be related to one's faith are important dimensions of the current debate. Physician-assisted suicide has been around for a long time and continues to be a challenge to religious and moral thinkers. The questions is how one's faith might be related to a physician's assistance in dying as a moral possibility. Should the moral community make every effort to abolish physician complicity with a patient's dying or death? Or should we both recognize its place as an exercise of mercy *in extremis* and regulate it so as to avoid the evil of indifference toward intolerable suffering?

Three chapters explore topics on the frontiers of medical science. "The Artificial Heart: Cyborgs and the Human Future" asks whether advanced medical technology is actually a threat to the future of humanity. A common concern is that technology may modify the human to such an extent that our humanity would be irretrievably compromised or that cyborgs will multiply to the extent that they become mortal enemies.

"Ethics and CTA" examines composite tissue allotransplants, including transplants of hands or perhaps of a face. The great debate is whether it is right to give anti-rejection medication that may kill the patient in order to transplant an organ that is, at best, only functional or aesthetic in value.

The embryo stem cell debate explores the fact that scientific possibilities for healing dread diseases or injuries have run up against strong religious objections. Should science have its curiosity curbed and its pursuit of the medical mandate stymied for the sake of preserving distinctively religious doctrines?

The abortion debate is examined in terms of what is thought "unthinkable" by a polarized citizenry. Case law from *Griswold* through *Casey* is brought into conversation with moral arguments advanced by groups favoring either choice or prohibition.

The final chapter deals with "Medicine and the Demonic." The widespread interest in Satan and demonic possession enlivens the imagination. People disagree over whether or not it contributes to new knowledge or perpetuates old superstitions.

Chapter 1

HUMAN SUFFERING AND THE MEDICAL MANDATE

"Pain is a more terrible lord of mankind than even death."
—Albert Schweitzer, 1931

"I say: Fear Not! Life still
Leaves human effort scope.
But, since life teems with ill,
Nurse no extravagant hope;
Because thou must not dream, thou need'st not then despair!"
—Matthew Arnold, "Empedocles on Etna"

"The practice of medicine, and its embodiment in the clinical encounter between a patient and a physician, is fundamentally a moral activity that arises from the imperative to care for patients and to alleviate suffering." Physicians have an "ethical obligation to place patients' welfare above their own self-interest and above obligations to other groups, and to advocate for their patients' welfare."
—AMA, *Code of Medical Ethics, 2006–2007 edition*

A thoughtful physician raised a fundamental question for medical ethics: "Is the purpose of medicine to relieve all human suffering?" His question actually posed two central issues: the nature of medicine as a moral enterprise, and the types of suffering to which physicians should respond. On that occasion, he was questioning whether physicians were obliged to attempt to cure all diseases and relieve all pain, but behind that question lies the enigma of the moral purpose of medicine itself. Honoring his question rightly demanded more than a simple yes or no.

Questions confronted in the clinical setting are first understood in light of the broader purposes to which medicine is committed. The issue of how medical science should be related to suffering is important not only to medical scientists, but to all those who work in areas related to health care, such as the pharmaceuticals and insurance companies. Is there, or should there be, a guiding vision that inspires and guides the work of those who claim to be involved in "health care"?

Another facet of the question pertains to the message being sent to those who are afflicted with serious disease or injury. What are the obligations of the healthy and skilled toward those suffering from disease or injury? Denying any obligation to relieve human suffering contributes to despair and humiliation on the part of those who are in the midst of assaults on the body or brain. Christopher Reeve's fall from a horse left him quadriplegic and requiring respiratory support. Michael J. Fox, Muhammad Ali, and over 500,000 Americans suffer from Parkinson's or other neurological injury or disease. Others suffer from exotic diseases that are at present untreatable and incurable. Those who suffer from diseases like progeria, for instance, are neglected both by medicine and the media. The reason typically given is that there are so few people affected by the disease that research is not cost effective; thus, no treatments are developed.

Emotionally and ethically, such stories are compelling since they force the issue as to reasons behind the continuing suffering of people with incurable disease. The answer is hardly simple. Complexity and complications seem to characterize any response. The popular and emotional response is typically a resounding and emphatic "yes—all human pain and suffering should be the object of medical research and the development of cures."

This chapter reflects on the question of medical science as a moral enterprise. What moral vision or commitment informs or inspires the medical profession? What commitments and perspectives are involved in becoming a physician and pledging that one will honor the profession and pursue its aims? What moral mandate should promote, guide, and inspire medical science? The insights of religion and the wisdom of science come together in public policy in ways that underscore the importance of the question and its answer.

Human Suffering and the Problem of Evil

Human suffering poses profoundly the problem of evil and how medicine relates to its effects on people and other creatures. The problem is as old as pain and reflections on its meanings in the human family. Thoughtful poets, philosophers, and theologians have struggled with the dilemma. The Christian tradition asserts both that God is perfect love and that God has absolute power. The problem of evil experienced from natural events seems a profound challenge to that central affirmation of faith.

Leibniz put it bluntly as a problem of theodicy, or the justification of God in the face of evil. Either God is absolute love, unable to remove or prevent evil, and thus not all-powerful, or God is all-powerful and thus creates or tolerates evil, which means God is unloving. Either the notion of power or that of love must be bartered if one is to believe in God. The grim reality is that suffering belongs to nature and human life. That fact is a challenge to thinking people who also insist on belief in God and experience God's love and power.

There is no fully satisfactory resolution of the problem on rational grounds. If there were an answer upon which all could agree, people would no longer be struggling with the issue in the halls of academia and in the clinical setting where diseases and injuries cause unspeakable levels of suffering and premature death.

Medical science has its origins in and has continued to develop as a moral response to human suffering. Physicians have responded to people who are diseased or injured. They offer specialty and personal skills to assist with cures where possible, but always with care. The entire range of people involved in health care join in the effort to mollify the effects of disease and injury and assist the person's return to wholeness and health.

People involved in health care know the depths and extent of the problem of human suffering, not as an abstract problem of intellectual interest, but as a human problem calling for a response that is both practical and moral. The response is practical in that it attempts to *do* something, not simply theorize or sympathize. But the action is moral in that the good of the person is pursued by one who cares about their well-being and has the skills that may actually assist with healing and wholeness.

Human suffering stems from both natural and moral evil. Natural evil refers to human sufferings caused by "natural" events, such as earthquakes, tidal waves, hurricanes, tornadoes, and volcanic eruptions. Drought, pestilence, pandemics, and floods would also be included. Keats spoke of "the giant agony of the world," referring to the chaos and violence of nature in its upheavals.[1] Most natural catastrophes are unpredictable to any great degree of precision, though we can know that nature will predictably act in

---

[1] John Keats, *The Fall of Hyperion* (New York: Harper, 1902) 185.

ways that are indifferent to the well-being or best interests of people. Such events cause widespread human suffering, from injury to disease and death. Volcanic eruptions still cause widespread injury and death. Vesuvius, Krakatoa, and Mt. St. Helens present us with images of the violence of nature, as do the tsunamis in Indonesia and the earthquakes in Turkey. Nature is beautiful in its bounty and seems friendly and benevolent when oceans are placid, the weather favorable to agriculture, and the winds calm.

But nature is not always beneficent. Tragedies associated with natural disasters are widespread and will continue. The hurricane that devastated New Orleans and the floods that ravaged the Northeast are reminders of the destructive forces of nature. The universe is unfinished. Giant upheavals throughout geological history have brought Earth to its present stage of geological development, but its journey in space and time has been marked by cataclysmic events, leading at times to the death of most living creatures. Dinosaurs lived for 150 million years but were apparently killed en masse by a catastrophic event of nature.

Such events are not evil in themselves; they are simply nature showing the effects of templates moving along one another, or the inferno at the heart of a planet where molten lava contributes to the forces of gravity, or the impact on the earth by colliding with a meteor. Earth's venture in space is made possible by forces only recently coming into the purview of science, but those forces are still beyond its ability to control. Since such disastrous events have no human origin, insurance companies like to refer to them as "acts of God." The aim, of course, is to dodge any responsibility for compensating victims. But the inference might be drawn to the effect that God is the source of such calamities—a direct challenge to the central notion of the goodness and love of God.

While natural evil stems from events in nature, moral evil relates to what people do to others that causes suffering and pain. The Apostle Paul spoke of "the mystery of iniquity"[2] or the "lawlessness" of people who defy or disobey the will of God and thus cause profound injury or death to people and other creatures. Slavery, torture, mutilation, genocide, and ethnic cleansing are evils invented and used by people against people. Genocide is still a cold reality that has caused the deaths of millions. Stories of such horrors can be found in Scripture, as in the holy war against the Amalekites.[3] Religious leaders blessed and demanded the practice in the name of God.

---

[2] 2 Thess 2:7.

[3] See the account in 1 Samuel 15. Apparently the prophet had more tolerance for carnage than the king.

Native Americans were reduced to a fraction of their populace with the movement to settle the West in the United States.

War is organized killing on a massive scale. Twenty million people were killed during World War II. People killed people by the millions in the major wars of the twentieth century, and the twenty-first century seems to be even more ruthless and perhaps headed toward even more extensive destruction. Weapons of mass destruction may now be used by the terrorist, the heartless dictator, the misguided president, or the unprincipled premier. All that is required is the technology of destruction, the means of delivery, and the human will to use them. The latter seems in abundant supply.

In all these ways and more, people are confronted with the profound question put so bluntly by William Faulkner: "If there is a God, why doesn't he do something?"[4] That moral reaction to human suffering is understandable as a reflection on the problem of evil in human suffering. Why is there AIDS or the virus that causes it? Why should Ebola or Marburg lurk in the hidden places and suddenly appear with an outbreak that kills ninety percent of all those infected? Tuberculosis, malaria, polio, and poverty are among the weapons of mass destruction ravaging areas of Africa, while developed nations have virtually eliminated such problems.

The devils in *Paradise Lost* sat on a hill and debated the questions of free will and fate and "found no end to wandering mazes lost." Milton despaired of finding any satisfactory or rational explanation to the problem of evil.[5] Medical science is motivated not by clever answers found in philosophical or theological reasoning, but in practical responses to the effects of disease on people. The great religions of the world affirm that God has done and is doing something about the problem of evil, not by eradicating the sources of the problem, but by inspiring responses that seek preventive remedies and cures for those who are directly affected by disease or injury. Regardless of how the calamity happened, God works through people who work with God to discover answers that are practical and personal, not abstract and ineffective.

---

[4] William Faulkner, *As I Lay Dying* (New York: Vintage International, 1930).

[5] John Milton, *Paradise Lost*, bk. 2, l. 555, ed. Thomas Newton (London: 1795).

Human Suffering and the Purposes of Medicine

The relief of human pain and suffering is typically taken as one of, if not the primary ends of medicine.[6] Medicine is a moral enterprise insofar as it works against natural and moral evil experienced in human suffering, but is all pain to be relieved, or is all suffering the object of the medical search for a cure? Stating the question begs for further distinctions to be made between pain and suffering and the appropriate goals of therapeutic medicine.

Pain and suffering are related, of course, but they can be distinguished. Pain refers to the response of the body to trauma that registers in the nervous system and brain. Pain is physical; suffering is psychic and emotional. Suffering is thus broader in scope than is the notion of pain. Suffering may include pain, but it goes beyond what may register in the neural pathways of the nervous system.[7]

One can suffer the trauma of disaster, the deep grief of loss, and the anxiety of threat without being in any pain associated with disease, illness, or injury. Poverty, social disruption, and political instability contribute to widespread human suffering, though medicine has neither a curative nor a palliative drug for social disruptions. The hungry and dispossessed suffer deprivation and grief that cry out for relief. Women worry about rape and contracting HIV in refugee camps. Infants contract HIV from infected mothers. The child dies from no fault of its own. Families grieve the loss by arrest and imprisonment of sons and fathers, many of whom will be tortured and killed.

The social and human dimensions of suffering were the subject of discussions among religious and political leaders from various parts of the world assembled at the Carter Center in Atlanta, Georgia. Christians, Jews, Muslims, Hindus, Buddhists, Ba'hais, and other religious leaders were represented among others in a discussion of the problem of suffering in the world. Former US President Jimmy Carter summarized the insights and message from the weeklong conference. He acknowledged that major differences of opinion predictably emerged in such a disparate group, but there was one goal on which they could all agree: "We should all be involved in the prevention and alleviation of suffering."[8]

---

[6] Eric J. Cassell, *The Nature of Suffering and the Goals of Medicine* (New York: Oxford University Press, 1991) 32. The book and his article below have the same title.

[7] D. Soelle, *Suffering*, trans. E. Kalin (Philadelphia: Fortress Press, 1975) 16.

[8] Quoted in *Time Magazine* (October 21, 1996) 54.

Carter's comment indicates that at least part of the problem in addressing the relation between the purposes of medicine and the problems associated with pain and suffering has to do with the different meanings of the terms and the central models or images by which the issue is addressed. The physician's perception of the issue is typically focused on those clinical cases in which they have dealt with patients in pain from disease or injury. The Carter Center's perspective focused on the violation of human rights, poverty and malnutrition, the effects of war, famine and pestilence, and the breakdown of social institutions from civil war or disruptions.

The variety of images that emerged in that conference helps to give breadth and depth to the question of medicine's moral mandate. No single image is adequate to grasp what is meant by pain and suffering, or to assess their relation to the moral purposes of medicine.

This variety of images of pain and suffering forms a backdrop for this discussion. The purposes and goals of medicine cannot be separated from human suffering in its multiple dimensions, for medicine is the science of the body and human relations in life, which are bodily in nature. People are embodied creatures who confront the world as identifiable and substantive selves. People live and have their being in multiple dimensions, and they are involved in complex human relations, but all these come together in the person as embodied. People may and do suffer in ways that involve physical pain, mental anguish, spiritual grief, and emotional distress.

The diseases or injuries that bring patients to the attention of healthcare workers are, to a considerable degree, evidence of the problem of evil in human life. Medical science is itself a human response to the experience of evil that tends to deprive people of health and happiness and leads to a premature and perhaps ugly death.

Medical Science and Its Purposes

Even so, there are at least three reasons to believe that science has no mandate to relieve *all* human suffering. The first is related to *the internal limits of medicine.* Medical science is a human enterprise and thus has finite capacities. Medicine is simply incapable of resolving all the factors that cause human misery, much less eliminating all pain and suffering experienced in the broader arena of social circumstance and political realities.

The limits of medicine pertain both to the ends of therapy[9] and to the capabilities of science. Many diseases cannot be cured. In a sense, science is simply incapable of eliminating all pain and suffering, and it would do well to recognize its limits, even while struggling against them. The beginning of wisdom is in self-understanding and coming to terms with the reasons for which energy and action are expended.

Medical science is committed to noble ends and ideals that both inspire great effort on behalf of others and present temptations to think too grandly of the human possibility. Physicians face the special temptation of "playing God," in the sense of overreaching or over-promising, believing that they can cure every disease or correct every injury, but medical science is neither omniscient nor omnipotent, as if all human problems might be resolved were the insights and techniques of the healing arts taken seriously enough. Such grandiose expectations reflect *hubris*, not truth.

The second reason pertains to *the possible corruption of science*. In an important sense, science *should not* attempt to relieve all human pain and suffering. There are some things that cannot be done morally.[10] Science ought not to assume it has a moral mandate to remove all pain and suffering. For science to assume otherwise would tend to legitimize whatever processes or technologies that might enhance the prospects for reaching the goal.

Enough is known about the narcotizing effects of certain drug and surgical interventions, for instance, that it is just possible, technically speaking, that science *could* remove all bodily pain and suffering. There may be ways of modifying people to make them impervious to pain and oblivious to suffering, whether their own or that of others.[11] A drug-induced stupor, for instance, might narcotize the pain, but not eliminate the torture that injures the body and assaults the mind. The brain might be modified by various surgeries in order to remove the capacities for caring or reflective, empathetic thought. In doing so, however, an important part of the *humanum* will have been severely compromised or lost. Science would be

---

[9] See Albert Jonsen, Mark Siegler, and William Winslade, *Clinical Ethics*, 6th ed. (New York: McGraw-Hill, 2006) 16, for a succinct list of the goals of clinical medicine.

[10] See R. Shattuck, *Forbidden Knowledge: From Prometheus to Pornography* (New York: St. Martin's Press, 1996), which explores five meanings of "forbidden knowledge," one of which is the impossibility of attainment.

[11] Apparently there are people who have no capacity for empathy or regret. They are capable of doing horrible things to others but show neither remorse nor guilt for their actions. Changes to the human psyche induced by drugs or genetic modifications might also cause such a state of mind.

working against the moral mandate to preserve health and restore capacities that are basic to human autonomy.

The third reason medical science has no mandate to relieve all pain and suffering is that *pain may be incidental to the purpose of human life and may serve the ends and goals of being human.* The purpose of medicine is correlated with the ends and purposes of life or the meanings and goals of human existence.

The patient is the primary ethical concern when asking questions pertaining to pain. Pain is not just pain; the moral question is derived from its meaning for the person in his or her particular situation in life. Many people endure significant levels of pain without it affecting their performance or their mental stamina. Some people have great capacities for enduring or even thriving under pain.

When totally absorbed by the game, for instance, an athlete can tolerate what at other times would be debilitating pain. The marathon runner must be able to overcome a body wracked by pain when the decisive moment comes as to whether one will endure for the finish line or drop out prematurely. The soldier can continue fighting with major injuries as long as the intensity of the conflict provides a concentration beyond the pain. Women can endure great pain in childbirth for the sake of the child and the family she intends.[12] The pain of childbirth can be severe but rewarding.

Pain and Suffering: Medicine and Human Anguish

A helpful distinction can thus be made between pain and suffering. What it means to be human is bound up in the human capacity for suffering, which may be closely related to but still quite different from pain. An ancient depiction of pain is portrayed as a penalty or punishment from the gods or a disruption of the intended order of creation.[13] The problem of pain is thus a major theme in the literature of world religions. Suffering poses both a central challenge to the belief in the goodness and power of God, and it may be an occasion for spiritual growth.[14] Some patients can relate their pain to the central stories or affirmations of faith, thus deriving meaning that brings an inward peace and relieves an agony that would otherwise be debilitating.

---

[12] See Cassell, *The Nature*, 35 where he shows that "the differences in methods used to control the pain…cannot be explained by the efficacy of relieving pain alone."

[13] S. van Hooft, "The Meanings of Suffering," *HCR* 28/5 (September/October 1998): 14.

[14] Cassell, *The Nature*, 35.

At least three meanings can thus be associated with pain and suffering. First, pain is primarily anatomic or physiological; it is the sensation of discomfort in some part of the body. Pain may be caused by traumatic injury, disease, or other disorder, but the sensation is transmitted through the nervous system. Second, pain or suffering may stem from distress, either mental or physical, caused by great anxiety, anguish, grief, or disappointment. Third, pain and suffering may relate to the enormous sacrifice of time, effort, and energy involved in order to accomplish a personal goal or perform an important task.[15]

Medical science is involved at all three levels, of course, though more direct significance is typically assigned to the anatomical meanings where medicine is concerned. If pain has to do with the body, science is not only interested but has a moral mandate relative to its control and management, but medicine has no moral mandate to eliminate all pain.

The Need for Pain. One obvious reason that all pain should not be eliminated is that the ability to perceive hurt or injury is the body's first line of warning that it is under assault. People born without any sensation of touch or the ability to feel pain have enormous problems avoiding serious injury. Not to be able to feel that one's hand is on a hot burner or plunged into boiling water, for instance, would result in a serious burn before the source of the injury is removed. Pain is the body's warning system, registering symptoms before there is serious and permanent damage.

A friend who recently underwent multiple bypass surgery had experienced weakness, but no pain. Since a long trip was planned, he decided to check in with his physician. Tests showed the occlusion of six arteries and veins, which required immediate intervention. He survived the bypass surgery and has regained energy and vitality. Doctors said his strong heart explained the lack of symptoms. As his experience shows, pain may call attention to an impending but preventable health crisis and thus have a positive purpose in the human experience.

Further, pain may be caused by physicians in treating disease and injury, as when bones are broken to set them properly. Surgery and other interventions will cause pain in order to create the conditions for healing to take place. Chemotherapy is certainly painful and causes its own type of suffering. Physicians are attempting to kill the cancer before the cancer kills the patient. Open-heart surgery is certainly painful, and the pain from the experience will not go away quickly, but creating pain is done in the therapeutic process of correcting conditions and thus restoring health. The

---

[15] *Webster's New World Dictionary*, 3rd ed. (Prentice Hall, 1991) 971.

benefits from the surgery typically far outweigh the burdens of pain and suffering.

The reality of suffering is in some ways a more complex subject than pain, though the two are interrelated and often interconnected. Cassell says suffering is a combination of grief and a sense of loss and regret that one's body may experience. He argues that "suffering occurs when an impending destruction of the person is perceived."[16] He sees suffering as a type of *angst* or profound anxiety that continues until the threat of disintegration has passed or until the integrity of the person can be restored in some other manner. He thus defines suffering as "the state of severe distress associated with events that threaten the intactness of the person."[17] Pain may trigger the onset of suffering, by his account, since the person may recognize or believe that his or her life is in jeopardy—that death and inevitable deterioration have set in.

Second, Cassell explores the meanings of suffering in the light of personal goals and purposes. The individual has multiple roles and lives out of a cultural context, all of which give definition to the pain and suffering one experiences. The person is a social, not simply an isolated, insular being. Suffering belongs to social existence, since all persons are involved in the body of humanity. As Hauerwas says, no two sufferings are the same.[18]

Third, Cassell notes that suffering can occur in relation to any aspect of one's life, whether in social roles, group identification, the relation with self, body or family, or the relation to a transpersonal, transcendent source of meaning."[19] He argues that suffering is related to what might be regarded as an existential threat; suffering occurs when one feels a threat or when one perceives that life is under siege.

That may be true existentially. But suffering may gain its greatest moral significance entirely apart from, or in spite of, any perceived threat to the self.

Beyond Pain to Transcendent Purpose

At this level, Cassell might be asked whether suffering is primarily a matter of sensing the disintegration of oneself. Or might suffering have dimensions

---

[16] Eric J. Cassell, "The Nature of Suffering and the Goals of Medicine," *NEJM* 306/11 (1992): 639–45.

[17] Ibid., 640.

[18] Stanley Hauerwas, *Naming the Silences: God, Medicine, and the Problem of Suffering* (Grand Rapids, MI: Eerdmans Publishing Co., 1990) 3.

[19] Ibid.

of meaning that have very little to do with one's own sense of disintegration or threat of death and relate far more powerfully to a sense that others are being subjected to serious abuse, injustice, or injury?

Two issues make a difference in the way that question is answered. The first pertains to what might be called reductionistic medicine, which treats pain as a symptom of something wrong with the person's body. That approach is likely to ignore or miss significant types or dimensions of pain. It is also likely to miss significant sources of the patient's distress that contribute to the presenting problem. The person/patient must be known in context before an accurate or adequate diagnosis of personal ills can be made. One experiences dis-ease *in context* whether it is familial, social, economic, religious, or political. That is why the patient's "history" is so vital to diagnostic procedures as well as to therapeutic interventions.

There is little value in dealing with high blood pressure as a physical symptom if the contextual reasons are overlooked as to why the pressure is out of control. A woman who presents with abdominal pain for which no apparent disease or obstruction can be found may be suffering from her family or social context. Her "pain" may be real but entirely unrelated to disease or physical injury. Her symptoms may be related to sexual abuse, which, if undetected, may remain a source of suffering as if she were ill. Such women are often subjected to exploratory but unnecessary surgery, which creates other physiological problems. She may wind up with extensive difficulties all because the source of suffering was not identified as the cause of the presenting pain.

A second recognition is that suffering may be embraced as an ennobling and redemptive response to injustices in the world. Embracing the pain of the world may be related to a transpersonal, transcendent source of meaning, as Cassell says. Evils in the world may be attacked in such a way that pain and suffering are unavoidable or embraced for themselves. For any number of social reformers, suffering directly from the social disruption might be entirely avoidable. But empathy with those who suffer "the slings and arrows of outrageous fortune" (Shakespeare, *Hamlet*, II, ii, 56) from social or political injustices drives one to act on behalf of the oppressed.[20] The activist has such an "identification with the problem" that suffering becomes inevitable. In that sense, suffering and pain are "invited" or accepted into a person's life. Instead of remaining aloof and detached from

---

[20] See Wendy Farley, *Tragic Vision and Divine Compassion: A Contemporary Theodicy* (Louisville: Westminster/John Knox, 1990) 69ff., for an exposition of the reasons for which one may become involved in what she calls compassionate resistance.

the human situation, one becomes involved and thus suffers with and on behalf of others. As Jürgen Moltmann puts it, "We begin to suffer from the conditions of our world if we begin to love the world. And we begin to love the world if we are able to discover hope for it. And we discover hope for this world if we hear the promise of a future which stands against frustration, transiency and death."[21]

Cassell seems to go too far in saying that such suffering is perceived as a threat to personal existence or a fear related to the deterioration of the self. Moltmann is describing a willingness to suffer on behalf of others but without any sense of self-annihilation. Indeed, in some profound sense, the individual is defining the self in and through the suffering that is accepted or invited as part of what Hauerwas calls one's moral project.[22]

Cassell comes close to this in his comment that "suffering is reduced when it can be located within a coherent set of meanings."[23] Saying that suffering is "reduced" may be misleading, however, since one's own suffering might actually increase as a consequence of moral commitments. German Christians like Dietrich Bonhoeffer[24] and Martin Niemöller discovered this reality when they chose to resist the Nazis for the sake of the Jews, but their suffering was morally necessary, even if politically and personally threatening. For such people, suffering is transformed and integrated into a developing and maturing spirituality in the light of personal or transcendent values.

For reasons like these, the person may be surprised to discover that others believe they are suffering. Mother Teresa was once told by a reporter that he would not work under such conditions as she did for a million dollars. Her response was immediate and telling: "And neither would I!"

Doctors Without Borders (Médecins Sans Frontières, or MSF) is an international group of physicians who go to the aid of people who have been through disasters such as earthquakes (Iran), epidemics (such as the AIDS pandemic in southern Africa), or war (as the insurgencies in Afghanistan, Iraq, and Dafur). MSF was awarded the Nobel Peace Prize in 1999 for their

---

[21] Jürgen Moltmann, *Religion, Revolution and the Future*, trans. D. Meeks (New York: Scribner's, 1969) 62. See also Moltmann's *The Experiment Hope*, ed. and trans. D. Meeks (Philadelphia: Fortress Press, 1975) esp. ch. 6.

[22] Stanley Hauerwas, "Reflections on Suffering, Death and Medicine," in *Suffering Presence: Theological Reflections on Medicine, the Mentally Handicapped, and the Church* (South Bend, IN: University of Notre Dame Press, 1986) 25.

[23] Cassell, "The Nature of Suffering," 644.

[24] See Dietrich Bonhoeffer, *Ethics*, ed. Eberhard Bethge (New York: The Macmillan Co., 1963) 200–201.

meritorious humanitarian service. The citation spoke of them as a "fearless and self-sacrificing helper (that) shows each victim a human face, stands for respect for that person's dignity, and is a source of hope for peace and reconciliation."[25] Six Belgian physicians with the organization were killed by rebels in Afghanistan. The doctors had gone in response to their sense of profound human need and their commitments to the moral mandate of medicine. In being willing to suffer with the suffering, they made the ultimate sacrifice.

For Christians, the paradigm is the story of Jesus' being "made perfect through suffering" (Heb 2:10), which, in turn, is a clue to the way in which Christians may invite and integrate suffering into their own life experience. What is at stake, however, is not the suffering of pain or the fear of death or personal disintegration. Their pain and suffering has distinctive characteristics. First, the suffering involved is not related to a simple coincidence of bad things happening to good people.[26] It has nothing to do with wandering viruses, genetic disease or painful accidents that wrack the body with pain and cause one to worry about a premature or even agonizing death.

Second, such suffering is related to the fact that the death of Christ delivers the person from a paralyzing *fear* of death, not from death itself. Death may become an inevitable or at least predictable outcome of the suffering associated with the moral effort or transcendent cause,[27] but death anxiety tends to immobilize people and thus undermine moral resolve, making them incapable of pursuing what they perceive as God's will in and for the world. The writer of Hebrews sees death anxiety as "a lifelong bondage" or an imprisonment of the spirit (cf. Heb 2:15). It is precisely the concern with self-preservation that runs counter to this posture toward the world and its ills. People may be delivered from self-centered pain avoidance to take on the suffering of the world. As they do so, as the writer of Hebrews

---

[25] From the Nobel Peace Prize for MSF documents, a copy of which is distributed by MSF.

[26] Harold S. Kushner, *When Bad Things Happen to Good People* (New York: Avon Books, 1981). Rabbi Kushner reflects on his son's progeria and how he deals with the problem of evil in the light of his religious beliefs and perspectives.

[27] Death is not sought as such nor as a means to a political or military goal. Nor is this approach to be mistaken as a type of suicide. The aim is not to become dead but to accept death should it happen as a consequence of forcing the moral issue. Neither suicide bombers nor Christian martyrs are deterred by the threat of being killed. But the aims and purposes, as well as the means and ends, of the two are diametrically opposed.

says, they also take up the suffering of Christ and become, like him, perfected for God's work in the world. The writer is talking about vicarious suffering, that is, suffering on behalf of others, which is basic to the development of character and is vital for the redemption of the world.

Such vicarious suffering has had an ennobling and transformative effect on the human family, and it has inspired and motivated some of the great social reform movements of the world, including but not limited to people like Mahatma Gandhi, Nelson Mandela, Martin Luther King Jr., Mother Teresa, Florence Nightingale, and Harriet Tubman. Both in their writings and their actions, such people have brought light to the debate and inspired noble actions on behalf of social justice.

The goal of medicine, therefore, is not the elimination of vicarious suffering. Suffering is morally desirable and ennobling when accepted on behalf of worthy goals and personal moral projects. Neither time nor space allows dealing with all the sources of human suffering and distinguish between those that seem to mandate interventions from medical science from those that do not. Broad categories can be defined, however, that help to clarify the moral mandate for medicine as it relates to human suffering.

Suffering and Injustice. Suffering that is caused by social or political injustice is not specifically related to the mandate for medical science, since it is largely beyond the capacities for medical cure. Physicians will nonetheless have a mandate to treat victims with whom they come in contact and to use their (considerable) moral influence to shape more humane social structures and practices. The near-universal condemnation of physicians who cooperated with the Nazi pogroms against the Jews and other minorities has shaped every code of medical ethics since the close of World War II. Medicine is now a post-Holocaust enterprise with the constant reminder of the moral distinction between using the powers of medicine to bless and not curse people. Medicine is not to become hostage to political ideology. The central task of medicine is to treat the patient as person, whether victimized by a wandering virus, the accidents of genetic mutations, or the vagaries of social or political movements.

Suffering associated with epidemics (HIV) or from genetic faults (Tay Sachs) often can and should be ameliorated or mitigated. Tragically, little attention is given to those afflicted by some diseases, such as progeria. The defense often given is because there are too few to justify the expenses in research and development of therapies. Science may never find cures for Alzheimer's or Parkinson's or certain other lethal afflictions, but it should never stop trying.

Judaism, Islam, Christianity, and other world religions fully support the effort to seek cures for all diseases. There is no line drawn between "optional" and "necessary" diseases to treat. The psalmist declared that the Lord "cures all our diseases and heals all our infirmities" (Ps 103:3), which implies that the medical mandate is tied to God's will for human wholeness and health. Jesus healed "all manner of diseases" and suggested that others would do even greater things (John 14:12). He seems to have anticipated and supported the wondrous developments in medical science witnessed in the recent past.

Suffering and Natural Disasters. There is no moral mandate, however, for science to prevent the suffering caused by tornadoes, hurricanes, tidal waves, earthquakes, or floods. Such natural disasters cause extensive pain and suffering, but the events of nature are not subject to human control. The moral mandate is to ameliorate their consequences for human health. Nations will organize politically to rebuild damaged or destroyed structures, rescue the homeless, and treat the injured, as they did in response to the tsunami in Bali and the earthquakes in Turkey. Physicians are obliged to assist the injured and diseased using the technologies and wisdom at their disposal.

Furthermore, many of the most threatening or impending disasters can be prevented. The moral challenge is often little less than the development of the political will. War, with all its attendant destructive forces, can and should be prevented. Significant steps in this direction have been taken with regard to weapons of mass destruction. People everywhere seem now able to breathe a bit easier since the demise of the Cold War. Treaties to lower and/or destroy large numbers of nuclear weapons seem to have pulled the world back from the brink of nuclear holocaust.

Considerable credit for the international attention given the threat of nuclear war should go to a group called Physicians for Social Responsibility. This group was devoted to influencing public policy and raising the awareness of the public to the dangers of threats such as nuclear war and widespread radioactivity from nuclear bombs or other devices.[28] The prospects of bacteriological warfare agents are also alarming to physicians. Agents that are genetically engineered pose a special threat. They are more

---

[28] See Helen Caldicott, *Missile Envy: The Arms Race & Nuclear War* (New York: William Morrow, 1984). Caldicott is a pediatrician and founding member of Physicians for Social Responsibility and founder of Women's Action for Nuclear Disarmament. She has lectured and written widely on the dangers of nuclear weapons and their threat to the future of humanity.

threatening than chemical weaponry since the diseases will spread rapidly through human populations and perhaps into the second and third generations before cures can be found or vaccines can be developed. Unfortunately, political and military objectives are often diametrically opposed to medical concerns for the health and well-being of people.

The fact that physicians get involved in sociopolitical activities to prevent widespread injury to people reflects ethical commitments basic to medical science. The principle of beneficence, or seeking the medical well-being of the patient, also requires doing no harm or preventing harm where possible. The best strategy toward weapons of mass destruction is to prevent their use for any reason.

Physicians are also active in campaigns involving high-risk behaviors that have such negative effects on patient health and well-being, not to mention their families and society at large. These involve efforts to reduce, if not eliminate, the use of tobacco, to use life-saving restraints in automobiles, helmets for motorcycles, and to enforce age-related limits for drivers of ATVs. The long-term effects of using tobacco products are extensively destructive to the coronary-pulmonary system, not to mention tobacco products' cancer-causing properties. Lifestyle choices are at issue in the use of seat belts or helmets for motorcycle riders. The bodily pain experienced by the spinal-cord injured or disease-ravaged is significant enough, but the suffering of those who bear the burdens associated with neurological injury is simply incalculable.

The will and commitment of physicians to the medical mandate will also be tested with epidemics that cause widespread death. Fears of epidemic infections contribute to hysteria and panic, one aspect of which is to withdraw into what are thought to be safe places. The instinct for self-preservation often suppresses the moral commitment to treat injured or infected patients. The early days of the AIDS epidemic saw physicians and other healthcare workers refusing to treat the infected. A chaplain who ministered to AIDS patients with touch and prayer set a nobler example. With better knowledge and the use of universal precautions, physicians now deal with such patients with a wiser courage. The threat or fear of a new outbreak of Avian flu (or Ebola, or Marburg, or any other of a variety of lethal agents) will test the resolve and moral commitments of the current generation of physicians. Dr. Rieux was honored in Camus's *The Plague* as a physician of courage and compassion who followed the medical mandate. He remained with the infected rather than run for cover to protect himself. In a time of plague, we are all vulnerable; there is no place to hide. We can either be faithful to the transcendent call to serve or cower from the fear of death.

Medicine and the Relief of Pain

The subject to this point has been primarily about suffering. The focus is not necessarily pain, except as it is incidental or coincidental to the purposes and activities of one's life. Pain can be integrated into one's sense of purpose or the meanings one gives to living. Insofar as pain is simply part of the story of one's being and acting meaningfully in the world, as in training for athletic competition, it is not morally significant. Pain that is inflicted on people as in torture, or uncontrolled in the clinical context of patients who are dying with pain, as in refusing to provide anesthesia, poses enormous ethical issues and requires a moral response, however.

The Problem of Pain. Pain is morally significant when it is debilitating and excruciating. There are degrees of pain that may contribute to the loss of one's will to live. Chronic pain may make life unbearable and undermine the pursuit of all other purposes. When that happens, the purposes of medicine and the problem of pain come together. Science has a moral mandate to relieve pain that is unrelated to a person's beliefs and purposes in life. Pain, to be acceptable morally or medically, must have a larger significance than as a symptom of a profound disease or injury to the body or brain.

The moral mandate of medical science can thus be more easily focused on the management of pain than the relief of all suffering. All the skills and techniques of science should be marshaled to control pain that prevents people from pursuing life plans or commitments that give meaning or depth to living. For the patient for whom pain is a personal affliction and/or a barrier to meaning, science has the responsibility to relieve the pain, insofar as that is possible. It is a choice of quality of life over life that is debilitated and purposeless because it is paralyzed or preoccupied by unbearable and meaningless pain. As the physician/philosopher Albert Schweitzer once observed, "Pain is a more terrible lord of mankind than even death."[29]

In England, the Brompton cocktail, a mixture of morphine and/or diamorphine and cocaine, was first used in the 1930s to control symptoms in patients with advanced disease.[30] Physicians used the drug on a regular, four-

---

[29] Albert Schweitzer, *On the Edge of the Primeval Forest*, trans. C. T. Campion (New York: Macmillan, 1948) 92.

[30] Cicely Saunders, "Foreword," in *The Oxford Textbook of Palliative Medicine* (Oxford: Oxford University Press, 1993) vii.

hour regimen. The opioids remain a standard part of pain control even while other drugs have been developed. A self-administered regimen is provided patients who are terminally ill with cancer. The patient releases sufficient amounts of medication to control the pain and thus make it possible to pursue meaningful activities such as tending roses and flower gardens. Being able to enjoy a few months caring for flowers is preferable to living perhaps a bit longer but with debilitating pain that forces one to live in a nursing home or hospital.

Medical science has a duty to relieve pain (1) when the pain is limiting or debilitating, (2) when it is unbearable and thus becomes a constant preoccupation, and (3) when it is inflicted on others through malice, deviance, or injustice. A fourth should be added to emphasize the obvious. Pain should be relieved insofar as possible during medical procedures that make pain both inevitable and unavoidable.

Childbirth pain, for instance, may be endured for reasons related to a woman's commitments and desires for family, as indicated above. Even so, women undergoing childbirth should not be denied anesthesia contrary to her wishes. If she prefers to make the pain more tolerable or entirely controlled, her wishes should not be denied based on the knowledge that women have for centuries had children without anesthesia. No generalization can be drawn from historical practices or circumstances and current medical standards of care. Analgesics are now available that our ancestors knew little or nothing about and thus had no choice but to endure the pain. Pain during childbirth is simply no longer necessary or desirable for most women. Withholding analgesics contradicts the standards of care for compassionate medicine. Such denials are especially reprehensible when funding issues control the decision.[31]

Hospice and End-of-life Care. The hospice movement grew out of the awareness that efforts to relieve pain cannot wait for the root causes of disease to be resolved. While research continues, an all-out effort must be made to control pain in the dying.[32] Hospice has institutionalized the practice of managing pain in the dying patient. Patients with incurable disease may decide against further or aggressive interventions and accept hospice care, whether in the home or hospital setting. The patient should

---

[31] See "Suit: Women denied anesthesia in childbirth," *USAToday*, Tuesday, June 16, 1998, 7A, in which seven women alleged analgesics were withheld from them because they could not come up with cash to supplement the state's MediCal insurance payment which the physicians thought to be inadequate.

[32] Saunders, "Foreword," v.

have no fear of experiencing dysfunctional pain in their final days. The message to patients who come under the influence of hospice, in the words of Dame Saunders, is that "you matter to the last moment of your life, and we will do all we can, not only to help you die peacefully, but to live until you die."[33]

Hospitals have also developed palliative care units, and specialists are available to ensure the proper administration of opioids or other drugs. Palliative medicine has come of age and is now standard practice in medical care.[34] No patient should go through surgery without appropriate anesthesia. The anesthesiologist who stole analgesics from patients in order to feed his addiction violated both the respect due a patient and the rule of "do no harm." Patients complained of excruciating pain on the operating table. One woman in spinal surgery said it was so painful she prayed for death. Patients were in pain from the surgery itself, not just the disease. They remained fully aware of and could feel the scalpel. The surgery was justified as a way to treat the disease, of course. But the lack of pain relief was a severe violation of basic norms of medical ethics.[35]

Stories like this underscore the ethical norm regarding the relief of pain that is now basic to standard medical practice. Doctors have a moral and legal responsibility to relieve pain. That responsibility extends even to the point of ending life itself. According to the principle of double effect, relieving pain is acceptable even if, inadvertently and unintentionally, respiratory arrest occurs. The US Supreme Court embraced this view in its review of issues pertaining to physician-assisted suicide. The court ruled that actually assisting a suicide would be legally vulnerable, but where the intention is to relieve pain that inadvertently results in death, there would be no legal fault.[36]

---

[33] Ibid., vii.

[34] See D. Doyle, G. W. C. Hanks, and N. McDonald, eds., *The Oxford Textbook of Palliative Medicine*, 2d ed. (Oxford: Oxford University Press, 1993). This impressive and important book describes both the protocol and the procedures for managing pain related to various illnesses.

[35] "Doctor who stole patients' painkillers gets prison," *The Courier-Journal* (Louisville KY), 26 February 1997, A6.

[36] *Vacco, Attorney General of New York et al v. Quill et al*, US Supreme Court, decided June 26, 1997, and *Washington et al v. Glucksberg et al*, US Supreme Court, No. 96–110, decided June 26, 1997.

The problem of physicians refusing or failing to ameliorate pain is, of course, not new.[37] The relief of pain has not always been understood as a clear mandate governing medical practice. As Cassell put it delicately, "The relief of [pain], it would appear, is considered one of the primary ends of medicine by patients and lay persons, but not by the medical profession."[38] Strong reservations have been held by physicians as to the wisdom of prescribing drugs to control pain, even in ordinary cases of patient care.

Patient pain and physician neglect were given critical attention in the SUPPORT study conducted in the United States between 1989 and 1993. The study amounted to a startling reminder that physicians did not pay sufficient attention to patient preferences regarding the control of pain or other elements of an advance directive.[39] Interview data from over 9,000 patients expected to live no more than six months on average was analyzed, as was data from five major teaching hospitals in five different states. Phase One of the study was undertaken prior to the Patient Self-Determination Act of 1989. Phase Two was designed to correct the perceived deficiencies in patient care that were identified in Phase One. Attempts were made to improve communication between patient and physician. Doctors were given written prognoses, provided with reports of patient beliefs and desires regarding end-of-life treatment, the type of pain experienced, and their desire for information. Nurses were especially trained to facilitate discussions between patients, their families, and/or surrogates and physicians.

A major finding was that advance directives have no clinically important effect concerning resuscitation among seriously ill patients, nor did they have any effect on the use of hospital resources for those who died during their first hospital stay after enrolling in SUPPORT. Furthermore, not one of the five categories in Phase One was appreciably affected when the follow-up was done four years later. There was no change in the timing of DNR orders, in physician-patient agreement about DNR orders, in the number of undesirable days, in the prevalence of pain, or in the resources consumed.

---

[37] R. M. Marks and E. J. Sachar, "Undertreatment of medical inpatients with narcotic analgesics," *Annals of Internal Medicine* 78 (1973): 173–81.

[38] Cassell, "The Nature of Suffering," 645.

[39] "Study to Understand Prognoses and Preferences for Outcomes and Risks of Treatment (SUPPORT)" *JAMA* 274/1 (November 22, 1995): 1591–1636. See also *Journal of Clinical Ethics* (Spring 1995): 23–30, and *Hastings Center Report* (November/December 1995), Spec. Suppl.

Physician practice patterns seemed to have changed very little during the era of the study. Even the awareness of the SUPPORT study did not register a positive impact. The only area of appreciable difference was in the influence of the SUPPORT nurses. Among the findings were that (1) only one in five (20.2%) patients had a written advance directive; (2) only about 6 percent of these directives were noted in patient charts; (3) doctors did not discuss medical information with patients or family members; (4) nearly half (46%) of all DNR orders were written only a day or two before death; (5) doctors did not know patients' wishes about treatment; (6) pain was a common element among the dying; (7) 38 percent of all patients spent at least ten days in ICU; and (8) the costs of care were unaffected by the presence of advance directives in patients who died during their initial hospitalization.

These patterns persisted in spite of the extensive discussions of treatment issues in the public sector. The Clinton administration's attempts at healthcare reform, the crusade of Dr. Kevorkian, the initiatives for assisted dying in Washington and Oregon, and the passage of the Patient Self-Determination Act all taken together seem to have had little effect on physician practice. The issues seem now to be at the center of concern by hospital ethics committees, which attempt to educate healthcare personnel about patient rights to refuse treatment and to have their pain controlled. Hopefully, the sensitivities of physicians to patient needs have improved since the close of the study.

Physicians who refused pain relief gave a variety of reasons for their actions, thus ostensibly providing moral legitimacy for withholding pain medications. First was the fear of patient addiction and, second, the fear of being thought too generous in providing prescriptions for controlled substances. As to the fear of addiction, the issue is moot when the patient is terminal. The dying body does not become addicted to medications.

What has been shown, however, is that the zealous diligence of the Drug Enforcement Agency (DEA) and peer review are significant factors in suppressing the supply of pain-relieving medication. Fears associated with law enforcement are largely overblown. Physicians must, of course, be able to document the appropriateness and frequency of prescription medications. Where that is done, the DEA has no interest in micromanaging physician practices. Peer review can certainly be a difficult matter since physicians are something of a self-regulating guild. The fact is that some physicians are not only adamantly opposed to the use of narcotics, but they take it on themselves to police other physicians' practice. The net result often seems a crusading effort at the patient's expense.

Other reasons for refusal to ameliorate pain include physician suspicion about patient pain. As one nephrologist put it bluntly, "How can I trust what my patients say when they say they are in pain but I cannot find something wrong?" Furthermore, drug abusers are adept in the ability to con physicians into providing unwarranted prescriptions, thus requiring vigilance by physicians and pharmacists, but that is a different problem than the experience of pain in the clinical setting, where pain can be diagnosed as part of the disease or injury and it requires relief in the name of good medicine and solid morals. Elie Wiesel once commented on the ongoing suffering by Jews in societies that tolerate anti-Semitism: "If a Jew says he is in pain, you listen!"[40] The point applies also to patients who complain of pain.

Another major reason for physician reluctance to relieve pain is the awareness of the fine line between controlling pain and causing respiratory arrest. The patient may die instead of just having the pain relieved. The fact that this "indirect effect" is not legally actionable is of little comfort to physicians who do not want their patients to die. They prefer to err on the side of caution, and that means allowing pain that the physician believes to be tolerable.

The pedagogy of pain is still another reason that pain may be tolerated by physicians for the "good" of the patient. Beliefs associated with what patients might learn during the suffering process have been used to justify allowing pain to persist. The notion is that the patient should learn through pain what had otherwise been missed in their spiritual pilgrimage. As one person said, "The final stages of an incurable illness can be a vast wasteland. But it need not be. It can be a vital period in one's life reconciling him to life and to death and giving him an interior peace."[41] A similar contention was made by the late Pope John Paul II when he refused aggressive care at the end of his life and advocated physicians not attempt to find cures for certain

---

[40] Elie Wiesel, *The Oath* (New York: Random House, 1973) 214. Wiesel repeated this point at his appearance at the University of Louisville, April 10, 2000.

[41] Norman St. John Stevas, "Euthanasia in England: The Growing Storm," *America* (May 2, 1970).

diseases. He wanted to make a witness for accepting pain as a type of what he called a "cruciform existence."[42]

But there are important ethical differences between knowing that growth *may* take place through pain and insisting that patients bear terrible pain because they need the spiritual growth. The Pope's acceptance of pain as a spiritual discipline is well and good since it was mandated by his own religious and personal values, but it is not acceptable medical practice to impose that attitude on others. Medical paternalism has no uglier face than that in which medication is withheld for altruistic or "spiritual" reasons. Imposing an opinion of one who is not in pain (physician) upon one who has requested relief for reasons consistent with their own values and religious commitments is both morally and medically intolerable. It is, in short, a type of religious imperialism entirely out of place in the practice of medicine.

Cassell recommends dealing with the relief of pain by assisting the patient to reinterpret and thus more successfully manage one's personal pain. "In the clinical context," he says, "pain and suffering can be relieved by making the source of the pain known, changing its meaning, and demonstrating that it can be controlled and that an end is in sight."[43] Those are important goals to bear in mind when dealing with patients who may be in need of such counseling, but when pain is unbearable and dysfunctional to the point of robbing the patient of reason and reflection, such counsel can add to one's suffering and thus extend and deepen, not allay, the pain or comfort the patient.

The distinction between pain and suffering thus may be used as a justification for withholding pain medication. The "pain as punishment" tradition in some religious teachings should not invade the medical setting. That notion was rejected by both Job and Jesus. Nor should knowing that people may be so committed to causes that they are able to bear intense pain be used as a pretext for denying pain medication in clinical practice. Perspectives on the meaning of pain and suffering are so profoundly personal that they ought not be imposed by one person upon another. That pain may be absorbed into or integrated with the larger commitments of life is true, but that *this* patient is in pain and desires relief are the clinical facts that should govern the decision on the part of the caring and responsible physician.

---

[42] See *The Courier-Journal*, 18 February 2005. See also S. Hauerwas, *Silence of the Lambs* (Grand Rapids, MI: Eerdmans Publishing Co., 1990), who deals with the problem of fatal illness among children as a "cruciform existence."

[43] Cassell, "The Nature of Suffering," 641.

Conclusions

The experience of pain and suffering presents problems that are philosophical/theological and medical in nature. The responsibilities of medicine are extensive and significant. They are also limited in important ways. There is a moral mandate to relieve pain that is dysfunctional or that is relievable by ordinary medical techniques or interventions, and when such relief is desired by the patient. On the other hand, science has no mandate to relieve *all* pain and suffering. The world would be poorer were it not for those moral heroes who embrace the suffering of others in order to bring some semblance of justice to the social and political process. The acceptance of suffering must be related to the patient's sense of purpose or moral aims. Suffering is not to be imposed or simply neglected by medical specialists who think the patient should bear pain for reasons that seem good to the physician, but are inconsistent with the patient's religious values or life-defining commitments.

Chapter 2

FALSE HOPE: FAITH AND HEALING IN THE ICU

Ah, but a man's reach must exceed his grasp,
Or what's a heaven for?
—Robert Browning, "Andrea del Sarto"

Hope is a vital component of human health and is basic to the healing process. Patients who are depressed or have lost interest in life or the future are not good candidates for surgery, for instance. A person's chances for surviving life-threatening illness or debilitating injury are lowered when symptoms of depression persist in the patient. People are also far less likely to survive extremely stressful events or what seem to be unremitting and devastating circumstances.[1] An optimistic or hopeful outlook is crucial to good outcomes in all these areas.

Even so, optimism often masks a false hope, which manifests itself in a variety of ways. False hope is the presence of an optimistic outlook that healing will take place in spite of a bleak medical prognosis and the advanced state of a lethal disease or devastating injury. Physicians may be convinced that further medical interventions would not be medically beneficial, but the patient and/or the surrogate is convinced that further treatment is required to bring about what they call "a miraculous turnaround." The presence of false hope regarding patient recovery often goes unacknowledged by those most involved in clinical settings, and it receives scant attention in both religious and secular literature. The emphasis usually falls on the positive effects that hope and faith might have on healing, but the persistent presence of false hope in the clinical setting merits attention.

This chapter explores the meanings of false hope, as well as its origins and manifestations in critical health situations, and suggests an ethical

[1] See V. Frankl, *Man's Search for Meaning*, rev. ed., trans. I. Lasch (New York: Simon & Schuster, 1962).

response to such phenomena. Special attention will be given to Christian sources of and expressions of hope and their influence in the clinical setting. Similar phenomena appear in other religions and could be explored with benefit. I focus on Christianity because it is the dominant religious voice in America and the faith tradition most influential on my own personal and professional development. My purpose is to sort out the grounds for the phenomenon of hope in the clinical setting and to attempt to discern the difference between hope that is sustaining and genuine and that which is futile and false.[2] Two stories help to focus the issue.

S. was forty-six when she died of cancer that had started with a melanoma on her left arm. The surgeon who removed it was relatively confident that all tumorous material had been removed. If not, he said, it would reappear in two years. Unfortunately, it reappeared; it was also metastatic. She had lymphoma. They tried every procedure. S. fought the disease, and her hope for a cure remained strong. She even applied to the NIH for a new experimental therapy. As her sister put it, "That was our last hope." She died just before she was to enter the trials.

H. was severely injured in a head-on collision. He was maintained in a nursing home for three-and-a-half years in spite of a very dim prognosis. He was finally declared to be in a persistent vegetative state—the condition given extensive media attention in the cases of Nancy Cruzan and Terri Schiavo. His wife decided to terminate the nutrition-hydration supporting his minimalist life, because, she said, "That is what he wanted. He did not want to be maintained in a way that has little if any possibility for the restoration of consciousness." She believed that any hope for his recovery was based on denial and false expectations about his medical and personal future.

Strong feelings and heated debates have been generated over just who entertains true or false hopes in situations like these. Disagreements often turn to acrimony with accusations ranging from bad religious faith to malice toward the dying on the part of those who believe treatment should be terminated.

---

[2] See Jerome Groopman, *The Anatomy of Hope: How People Prevail in the Face of Illness* (New York: Random House, 2004) ch. 2, who describes the way in which doctors often instill false hope in patients with terminal illness.

The Paradox: Hopes True and False

That Christians are to live by hope is a given of the faith. Hope is a thoroughgoing motif in Scripture. Jürgen Moltmann has observed that even the major sections of the canon end with a view of what is to come. The very structure of the Bible is anticipatory or future-oriented.[3] Paul captured the inner dynamics of Christianity with a trinity of grace: "So now abides faith, hope and love" (1 Cor 13:13), he declared. And the writer of Hebrews treated hope as the dynamic of history: "Faith is the assurance of things hoped for, the conviction of things not seen" (Heb 11:1). Hope is a lively expectation of the unpredictable; its anticipations are rooted in faith.

People also embrace false hopes. These seem to be the "other side" of both good faith and solid hope. The phenomenon of hope is basic to living; the phenomenon of false hope seems its perpetual twin. To put it another way, people live by hope whether it is true or false, whether it is grounded in reality or based on fanciful thinking. False hope is wishful thinking or daydreaming in a way that denies reality. It is a way of construing one's future alternatives in a way that denies or refuses to assess entirely predictable but unpleasant outcomes.

What teacher has not listened to the youngster who speaks confidently about the high performance level he or she expects on an upcoming exam? To hear the student tell it, all is well, the exam will be relatively simple, and the test grade will show superior marks! The teacher's wry smile betrays a telling insight. The student has not kept up with homework, does not have good study habits, shows no other indications of having mastered the material, and, on a daily basis, does not show even a passing performance level. The student's "hope" is stated emphatically and even enthusiastically; there is an air of confidence that would otherwise be convincing, but the teacher knows otherwise. It may even be that the student is religious and expresses confidence in the future in religious terms: a miracle will take place, or God will give the student superior insights and knowledge so as to do exceptionally well on the exam.

The teacher's skepticism is neither a faulty religious piety nor a lack of belief in miracles. Nevertheless, skepticism wins out. The teacher knows that the student is denying reality and expressing hopes as a kind of fingers-crossed approach to the coming examination. Flunking the exam and not

---

[3] Jürgen Moltmann, *Theology of Hope*, trans. J. W. Leitch (New York: Harper & Row, 1967).

getting superior marks is entirely predictable. The future outcome seems certain.

The parallel between the classroom and the clinical setting is not difficult to make. Physicians and other health professionals have seen optimistic patients with an unmeasured confidence in positive outcomes even when the prognosis is terribly bleak. Retreating into a false hope and optimistic view of the future is one way of coping with traumatic news, of course. Without the safety valve of such a coping mechanism, one's emotional, spiritual, and even physiological systems might fail.

False Hopes as Looking for a Miracle

Beyond expressing such hopes as a coping mechanism, however, patients often "expect a miracle" as a matter of religious faith. For them, persistence in treatment is a way of being faithful to God.[4] The parents of a dying child might believe that the child's only hope is their unwavering faith that God will heal the child. They fear that God will not heal their child if they fail the test of faithful waiting. Presenting themselves as hopeful and faithful believers becomes necessary to prove their faith to those around them. The child will get well; this illness will be defeated; God will give them a miracle. Of these they are sure.

Alexander Pope once observed, "Hope springs eternal in the human breast."[5] He noticed that hope pervaded various dimensions of the human condition: the struggle with illness, efforts to restore life after natural disasters, and the enduring effort to escape poverty, but Pope was skeptical of hope as a way to health or material prosperity. As he put it, "Man never is, but always to be blessed. / The soul, uneasy, and confined from home, / Rests and expatiates in a life to come." As Pope noted, life seems to become more tolerable if not more blessed by the capacity for hope, even if it is a false hope.

Further, false hopes and genuine hope have a way of becoming confused in the human psyche. The paradox belongs to being human. People are mortals and thus incapable of perfectly envisioning either future outcomes or the actions of God. They can be motivated by hope even when, as so often happens, those hopes are grounded in wishful thinking. Exaggerated expectations that take the form of hope may set one up for

---

[4] Cindy H. Rushton and Kathleen Russell, "The Language of Miracles: Ethical Challenges," *Pediatric Nursing* 22/1 (January/February 1996): 65.

[5] Alexander Pope, *An Essay on Man*, in *The Complete Poetical Works of Alexander Pope*, ed. N. W. Boynton (Boston: Houghton Mifflin Co., 1903) epistle 1, l. 95.

disappointments that may, in turn, be devastating. Bitterness and anger may result and may be directed at the self, loved ones, God, the physician, or an unjust universe.

Scripture, Story, and False Hope

The paradox of the intermingling of true with false hope is also found in the literature that helps to shape the Christian view of the future and the beneficent actions of God. Scripture is replete with stories that engender hope; other stories make it obvious that hopes are often unrealistic. The stories of Daniel in the lion's den, the resuscitation of Lazarus, and numerous others have rightly pointed to the beneficent action of God on behalf of those who received divine favor. Such stories have also tended to engender expectations that combine rather bizarre notions of how God will deliver people from early death or devastating injury.

Some people engage in high-risk adventures, believing God will deliver them from harm, no matter how dangerous the action. Jesus himself was tempted to act in self-destructive ways based on false hopes, as the story of the temptations in the wilderness suggests (Matt 4:1–10). Fortunately, as Christian sources insist, he resisted the temptation and did so without sin (Heb 4:15). Jesus provides the believer grounds for distinguishing between the lure and promise of false hopes and the strength acquired from hope that truly reflects the promise of God.

The stories of miraculous healings in the Scriptures are another source of human hope. They often enliven a Christian's expectations for positive outcomes in the clinical setting. A "miracle" is considered a contradiction of scientific laws or an action beyond human explanations or technological capacities. Medical literature also contains stories about "remarkable recoveries" that defy all clinical explanations.[6] Knowing that the inexplicable might take place and that unpredictable recoveries have happened encourages the hope for a miracle even against enormous odds.

False Hopes and Medical Uncertainty

Another encouragement to hope is in the uncertainty of the medical data. *Probable* outcomes are different than *certain* outcomes or results. The book

---

[6] See, for instance, C. Hirshberg and M. I. Barasch, *Remarkable Recovery* (New York: Riverhead Books, 1995); H. Benson, *Timeless Healing* (New York: Simon & Schuster, 1996); and Larry Dossey, *Healing Words* (San Francisco: Harper, 1993).

One in a Million[7] recounts the author's experience with his wife, Jackie, who had a severe stroke that left her in a coma. Physicians said her chances for recovery were "one in a million." Six weeks after her health trauma, her husband filed papers for court permission to disconnect her from ventilatory support. The judge took the petition under advisement. During the delay to decide the issue, she had a remarkable recovery. A high school friend visited her in the hospital, took her by the hand, and called her by a nickname. She opened her eyes and spoke. Jackie had recovered consciousness and returned to the routines of life and a tour to discuss her experience, howbeit with obvious limitations from the stroke.

Stories like that tend to enforce exaggerated expectations in the intensive care unit (ICU). But should they? Likely, six weeks was not long enough a trial period to discern the long-term consequences to Jackie from the neurological trauma she had experienced. Thus, the petition to stop treatment was perhaps both premature and ill-advised. But the positive outcome will encourage any number of people to expect such outcomes even years after the bad news is rather settled. In the case of Hugh Finn, for instance, many in his family of origin still hoped for his recovery well into the fourth year of his extensive brain damage and persistent vegetative state.

False Hopes and High Technology

Another factor in false hope is the extraordinarily high level of expectations on the part of the American public regarding medical science. Myth and reality tend to merge and become indistinguishable when public-relations hype advertises the extraordinary achievements of medical breakthroughs or accomplishments. A miracle or cure in one area tends to generate strong expectations for miraculous cures in other areas of medicine. This secular equivalent of religious hope may contribute to excitement and psychic ebullience in the ICU. Should the physician resist such enthusiasm or attempt to introduce a note of reality into patient assessments of the medical condition, even more denial may be generated. The patient's "hope" may take the form of disenchantment and anger against the attending or healthcare team. *Hubris* gets mixed with physical pain and psychic shock as people come to terms with human mortality and imminent death. Americans do not take defeat easily or as a matter of course; it seems always someone else's fault, and physicians are responsible to fix or correct it. A failure to correct the problem simply exacerbates the anger that may well be directed

[7] H. Cole with M. M. Jablow, *One in a Million* (Boston: Little, Brown & Co., 1990).

at anyone who seems to wield power but refuses to do so: the physician, the healthcare team, the family, oneself, or God, the ultimate fixer.

Hope in Three Modes

Hope has importance in discussions about clinical ethics at three important levels. First is the pursuit of a near-term goal of healing; second, the pursuit of long-term goals of new medical breakthroughs; third, the transcendent vision of life beyond mortal existence. All three dimensions need to be borne in mind and brought into focus when dealing with hope and healing *in extremis*, that is, when there seems no ordinary medical reason to expect the patient to recover health.

Life-threatening illness has its own way of generating false hope, of course. The threat or fear of death has a way of wonderfully concentrating the mind, as Dostoevski observed. The short-term goal becomes the hope for restoration of function or other goals of medicine that seem to promise optimal health.[8] In this framework, the hope may be based on a realistic assessment of prospects for positive medical outcomes, or it may be based on denials of the progress of the disease, the impact on the physiological functions of the human body, and the poor outcomes that are almost certainly, that is, are probably,[9] going to take place. The combination of religious hopes for miraculous intervention plus the tendency to deny bad news often creates the mindset for hoping in spite of all evidence to the contrary.

One reality check for Christian believers is to recognize that God has never promised that all diseases would be cured in this life. Jesus often refused to heal persons he encountered, and when Paul spoke of the absence of disease or a body of perfection, he was speaking of the afterlife (cf. 1 Cor 15). The end or goal of human life is not perfection of the mortal body, but perfection *in God*. People are mortal and thus destined to die (Heb 9:27). Furthermore, the dying process may be lengthy and painful. Deterioration spreads throughout the body as one organ after another fails. The body is in the process of returning to the dust from which it came (Gen 3:16).

Even so, there are two senses in which God has promised the curing of disease. One is in the progress made possible in a divine-human covenant

---

[8] See Albert Jonsen, Mark Siegler, and William Winslade, *Clinical Ethics*, 6th ed. (New York: McGraw-Hill, 2006) 16, for a list of eight goals or goods of medicine.

[9] Probability theory is a type of calculation for making prognoses for outcomes given a particular disease and modalities of intervention. Based on statistical averages, a particular disease can be calculated as having a certain probable outcome.

that discovers and makes available cures for many of the maladies that afflict the human family. The second is in the eschatological future in which God makes all things new. A Christian perspective holds these two in tension as partners in a vision of the possible and the future promised by God.

The prospect for cure is consistent with Jesus' promise that we would do "greater works than these" (John 14:12). His comment was apparently an allusion to his miracles of healing, which so inspired and enlivened the hopes of the disciples. Jesus seems to anticipate the far more awe-inspiring things that would be accomplished by those who came after him.

A great deal of that promise has been realized through medical science. Amazing breakthroughs in the treatment of disease have taken place that simply stagger the imagination. Polio, smallpox, the plague, tuberculosis—the list could go on and on with the impressive victories of science over the diseases that maim our bodies and claim our lives for premature death.[10] It is not impious to say that these are "greater" miracles than Jesus ever did. He cured many people, but modern science has cured or prevented illness for millions of people. Jesus pointed the way toward what could be achieved when people apply their creative energies to problems once thought overwhelming and unconquerable. Such breakthroughs in medical science seem ample evidence of the relation of hope to human healing. First, we can affirm the relation of science and religion as covenants of faith and hope. Second, medical science in general can be seen as a human enterprise in the pursuit of the future promise of God.

But nowhere does the Scripture promise that all diseases or limitations will be cured. There is an element of nature that is both unfathomable and unconquerable for the human mind. No matter how clever the human manipulations of nature or how incredible the new insights that might be discovered or the technologies that might be invented, nature will not entirely bow to human know how. There are things that people will never be able to do, and that includes conquering certain diseases.[11] The occasional outbreak of Ebola and Marburg reminds us of what is apparently an extensive variety in the expression and forms of disease-inducing viruses.

---

[10] Daniel Callahan, "Limiting Health Care for the Old?" *The Nation* (August 15, 1987), suggests a "premature death" is one in which a person as not lived long enough to experience life's prime benefits and opportunities—children, satisfying work, and pursuit of intellectual goods, and valued friends. Also see his *Setting Limits: Medical Goals in an Aging Society* (New York: Simon & Schuster, 1987).

[11] See Daniel Callahan, *False Hopes: Why America's Quest for Perfect Health Is a Recipe for Failure* (New York: Simon & Schuster, 1998), who cautions against thinking science can totally conquer nature.

They also remind us of death by means or sources that are not yet obvious. There are challenges awaiting every new generation of medical scientists.

Furthermore, not everyone will be cured of disease. Some will die even from diseases over which science typically has mastery. People still die while medical specialists are baffled as to why. At times, there is simply no apparent medical or physiological reason to explain why a person dies, just as some people get well under circumstances that defy explanation.

Futile Treatment and Hoping for a Miracle

From a Christian faith perspective, two things seem rather obvious about the hope for miracle. First, it is entirely understandable that such hopes surround our dying loved ones. Second, the hope for miracle may be terribly misguided and groundless. There is a point at which the hope for miracle becomes a false hope, a hope against hope without any grounding in reasonable expectation.

The stories are all too numerous in the clinical context. People require sustaining interventions that are terribly costly in financial, personal, and human terms, but of little, if any, benefit to the patient in medical terms. The social, personal, and economic "costs" associated with stories like those of Terri Schiavo, Karen Quinlan, Nancy Cruzan, Sue DeGrella, and Hugh Finn (and thousands of others in similar circumstances in US health facilities) are simply incalculable. The problem is that such interventions yield no benefits beyond whatever emotional values there may be to people who insist on extending supportive or even aggressive interventions for reasons that seem almost entirely eccentric to healthcare professionals.

Physicians speak of such cases as futile treatment, or a medical intervention that either offers no physiological benefit or creates more burdens than benefits for the patient.[12] Regardless of what is done, the patient is not going to recover. Physiological or neurological function will not be restored.[13] Family members may insist on treatment, however, based

---

[12] See Joanne Lynn and James Childress, "Must Patients Always Be Given Food and water?" *Hastings Center Report* 13/5 (October 1983).

[13] See Jonsen, Siegler, and Winslade, *Clinical Ethics*, 24; and Lawrence Schneiderman, Nancy Jecker, and Albert Jonsen, "Medical Futility: Its Meaning and Ethical Implications," *Annals of Internal Medicine* (1990) 112:949–54, who speak of two types of futility: quantitative, which refers to probability that medical intervention will have benefit is extremely small—less than 1 in 100 chance; and qualitative, when there is extremely poor quality of life associated with medical intervention. The burdens imposed by intervention may outweigh any possible benefits.

on false hopes for patient recovery. Paul Ramsey calls it "hoping on in 'faith' when all hope is gone."[14] Such false hopes take a variety of forms.

One type of false hope is for a medical breakthrough *in the immediate future* that will benefit the patient. "What if they find a cure in the next few days?" is often heard in such contexts, as it was in deliberations about Hugh Finn and Nancy Cruzan. In cases where the patient is already moribund (about to die) or terminal (likely will die within six months), however, even a breakthrough announced tomorrow will likely be of no benefit. Some patients who are ill are beyond the prospects of healing. Multiple organ failure or the devastating impact on the brain may make all such expressions of hope or "faith in God for a miracle" a matter of hoping in nothing but hope itself.

Such "rescue religion" fosters a hope that feeds upon itself.[15] Those who are dying of AIDS may have no hope of healing in their mortal bodies. The disease may be so advanced that even a breakthrough that promises miracle cures or the discovery of a vaccine that promises to prevent this dread disease may come too late to do the dying any good.

Religious piety may also generate another type of false hope. Some people of faith believe it an absolute duty to sustain "life" without regard to cost or other considerations. They may hope for a miracle, or they may see the suffering as a test from God. Another expression of such hope is to say that the patient must be sustained until *God* causes or brings death about. Anything short of "doing everything we can" to prevent death is condemned as "playing God."

Regardless of the particular expression, hope is seen as integral to one's fidelity to God or vital to being a person of faith. Helga Wanglie, for instance, was an eighty-seven-year-old retired teacher in Minnesota who was diagnosed as being in persistent vegetative state. Her husband insisted on full code for Helga, however, and for maintaining her with every means possible. Physicians were unanimous in their assessment of the futility of such care. Even so, Mr. Wanglie insisted that his desire for sustaining her was a matter of religious faith. He embraced the notion of the sanctity of life as an article of faith and thus wanted her maintained as long as technologically possible. He believed strongly that only God should determine the time of death.

---

[14] Paul Ramsey, *Fabricated Man: The Ethics of Genetic Control* (New Haven: Yale University, 1970) 29.

[15] R. B. Conners and M. J. Smith, "Religious Insistence on Medical Treatment," *HCR* 26/4 (1996): 28.

Two problems are noticeable from a Christian perspective. One is whether the Christian hope can be so reduced to a biological minimum. That is, the Christian hope does not focus simply on the minimalist signs of life that are (still) present in the near-dead body. When the hope for life is so minimally construed, the person as responsive to God and others is lost in the expansive world of fantasy and misguided belief.

The other problem is whether there is not a serious error at stake in confusing a hope for healing with a transcendent hope. The ultimate hope for people of faith is rooted in transcendence. Their life in God is not to come to an end, but their life on earth may be limited to a short future. The hope for the afterlife is an important reason people of faith do not lapse into the false hope that absolutizes mortal existence. As the Apostle Paul put it, "If for this life only we have hope, we are of all people most to be pitied" (1 Cor 15:19). The Christian hope is not for the cure of all diseases or that every person will be restored to health, but that we will be brought to completion in God's own eternity.

A strong faith in the afterlife is one of the primary reasons certain patients refuse medical treatments when the prognosis is so bleak. They do not see the issue as one of lacking either hope or faith. They see it as a matter of a hope that sustains them during the dying process. The goal of living in faith cannot be reduced to a long life or even "just a few more days of life," while the impulses of faith and hope are sustained. If that were the case, Jesus' life fell far short of what God promises. Jesus died when he was thirty-three. He would have chosen to live a safer way than confronting and challenging the authorities had longevity been his goal. The reason for living is more important from a religious perspective than is how long one lives.

Hope and the Future of Healing

Perhaps the most important dimension of hope is related to the quest for new cures and new medical breakthroughs. In Christian thought, eschatology is a way of intending and pursuing the future based on the promises of God. Christians have experienced the down payment on what is yet to come. They do not simply wait for the future of God's blessing, but actively work toward its realization. The future is open; it is malleable, not static. Thus, planning and hard work are basic to hoping and intending the future. Christians are builders of the very kingdom for which they wait.[16]

Hope energizes the human search for cures. Centuries of medical science have brought about an extensive array of therapeutic interventions. Hope is the ground and reason for any medical science at all. Such projects are not pursued in vain, but with the promise that many good results will come from a process of trial and error.

We can thus say with reasonable assurance that many people will be cured of AIDS. Such confidence is based on the promised future of God. AIDS will someday be looked upon as the disease that wasted and killed millions but does so no more. It will occupy a place in memory and history much like that of smallpox and polio. All these killed their millions but no longer ravage the human family. Now there are no tears, no death from such dread diseases of the past.

Thus, the "miracle" of a "cure" is a matter of perspective. Those who have never had polio can thank God for the vaccine that "cured" them from that paralyzing condition even before they were afflicted. Prevention is a proleptic cure—a cure before the disease is contracted.

Clarifying the Nature and Focus of Hope

Theologians and ethicists thus may play a vital role in clarifying the nature of Christian hope and its relation to the clinical context. The place to begin may be to note that hope has a variety of expressions. Just which hope is realistic or promised of God is important as the relation of hope to healing is considered.

For those who are already moribund, or far along in the dying process, there is only the hope of eternity, not the hope for bodily cure, regardless of medical breakthroughs in the very near future. The body has already begun its irreversible process of dying. Seldom is dying a matter of there being only one source of illness. The body is overwhelmed, and medical science is and will remain powerless to intervene except to offer comfort care. That comfort takes the form of pain relief and the presence of persons who are important to the dying. That significant others are present fulfills the hope that we shall not be abandoned in our dying—that our importance to the living transcends our infirmity and uselessness. The loss of the ability to reciprocate should not presage the loss of our value to others.

---

[16] See Moltmann, *Theology of Hope*, 337; see also Moltmann, *The Church in the Power of the Spirit*, trans. Margaret Kohl (San Francisco: Harper & Row, 1977) 376f.

The dying person's hope is also that others shall not have to waste away in a similar manner. Their prayer is not only for themselves, but for the well-being of the other. The hope is that the cures that do not benefit this one will nonetheless benefit others who are yet to come this way. One woman, dying an ugly death from the ravages of cancer, said to her pastor that she wished she could take all the pain of the world with her so that no one else would ever have to know what she had experienced.

Specialist Knowledge and Medical Indications

The role of the medical specialist is crucial in assisting families to avoid false hope in the ICU. Solid medical data are basic to knowing that hoping for a cure may be little more than wishful thinking. Families are understandably anxious about loved ones who are near death or who have experienced severe medical trauma. They cling to every word of the physician, listening for some hint that recovery might be likely or even possible. The problem for the physician is often the tension between providing comfort and the ethics of truth-telling.

The need to confront the family openly and honestly about bad news is underscored at this point. The temptation may be to offer consolation by holding out some hope in the face of adverse findings. The family needs to be informed with solid medical data that leaves no room for wondering how the physician assesses the patient's condition.

Further, the medical staff needs to be consistent in communication with family. If any staff person, whether resident, student, or consultant, wavers or adopts a "comfort" posture or hopeful attitude, refusing to deal with the bad news, division and/or confusion will be generated in the family group. False hopes will thus be encouraged if not generated by someone on the medical team itself.

Family members of patients in persistent vegetative state, for instance, may strongly deny the tragic state of the brain. If they happen also to embrace the twin notions of the sanctity of life and the absolute evil of death, they may be convinced that everything possible should be done. The very appearance of open eyes and eye movement convinces the one who wants to believe that the patient has some degree of consciousness. The hope that recovery is still possible becomes virtually unshakable. Certain members of Hugh Finn's and Terri Schiavo's family, for instance, were convinced that Hugh or Terri recognized them during visits. Denial often takes the form of wishful thinking that, in turn, requires very little evidence to be sustained.

Getting unanimity among professionals can also be a problem. Disagreements may be provoked by different readings of the medical indications. Physicians are also subject to the beliefs and value commitments that object to the very notion of futility. Their "optimism" contributes to false hopes or the illusion of good prospects for recovery.

An eight-two-year-old nun was admitted to a hospital with myocardial infarction. A longtime friend and Sister brought along a lengthy statement of the nun's wishes not to have aggressive treatment. She had lived a long and fulfilling life and was now "eager and ready to go to be with the Lord," as she said in her living will. She wanted a peaceful death, unhindered by aggressive interventions. The attending indicated on her chart that there was to be no renal dialysis. She was to be allowed to die a natural death.

But a specialist called to review the case disagreed. He felt the issue was not entirely clear and that the nun's written statement did not settle the matter. He strongly insisted that renal dialysis would provide a totally different picture! He brusquely took over from the attending and, in a fashion nurses called intimidating, proceeded to order the placement of the shunt to proceed with dialysis. The nun died before the procedure could be initiated. Her heart simply could not take the additional trauma to her body.

A sixty-year-old man being treated for acute leukemia had been through several rounds of chemotherapy in an effort to induce remission. His white blood cell count was extremely low. The oncologist then injected colony stimulating factor, hoping to increase the white blood count, saying, "He just might respond." A primary care physician colleague was amazed that such further treatments would be imposed and declared, "If he is alive a week, it will be a surprise to me." The man died the next morning.

Futility and the Faithful Family

A second major source of conflict arises from within the family who is determined to be faithful to their dying loved one. Certain members of the family may promote the type of "faith" that is adamantly opposed to discontinuing treatment for a patient in spite of the nearness of death.

A seventy-seven year-old female nursing home resident was presented to a local hospital. Her diagnosis included urinary tract infection and left lower lobe pneumonia. Her symptoms included cough, congestion, low-grade fever, and wheezing. She had a history of severe dementia, Alzheimer's, hypertension, cerebrovascular accident, pulmonary embolus, multiple hospitalizations for urinary tract infections, pneumonia, and chronic renal failure.

At the family's insistence, the patient was placed on full code. She was treated aggressively for fourteen days. Tube feeding, renal dialysis, and pleural effusion for respiratory distress were employed. Even after she went into multisystem organ failure (brain, kidneys, lungs), the family still insisted on full code.

The medical staff requested a consult with the ethics committee. The physicians attempted gently and sympathetically to convince the family of the futility of continued aggressive treatment. The husband finally decided for a do not resuscitate (DNR) order, which the daughters reluctantly accepted. Two weeks later, difficult interactions erupted between physicians and family. Physicians felt strongly that continued high-tech care in the ICU was medically and morally unjustifiable. The family agreed to hospice. Medications, gastric tube feedings, and comfort measures were continued until the patient irretrievably expired. The crisis in medical care for this patient was a spiritual and ethical matter, but the problem arose more from the family than the patient or physicians. The physicians were unanimous in believing they were being asked to provide treatments that had no positive medical benefit. Nothing could be done medically that would have any reasonable assurance of restoring even minimal functioning for the patient.

Jonsen, Siegler, and Winslade[17] speak of "contextual factors" that figure prominently in decisions about patient care. The crisis comes not from disagreement about the clinical data, nor from an assessment of quality-of-life factors, nor even from patient preferences. Analyzing the case from these three areas is quite clean ethically. The conflict emerged with the demands of the family to "do everything possible," whether or not it seemed to clinicians as either medically or ethically mandated. Ethically, say the authors, there is no obligation to provide aggressive treatment for a moribund patient.

The "family" in this case, of course, was the immediate family of husband and daughters, a relatively small group of people. The larger the "family" the more complicated and difficult decision-making may become. The simple reason is that the larger the group, the more likely a considerable difference of opinion regarding appropriate or necessary care will be found. Many of the most controversial cases in medical ethics have been prompted by political action groups that attempt to become surrogate decision-makers for patients, especially when they disagree with decisions on ideological or religious grounds.

---

[17] Jonsen, Siegler, and Winslade, *Clinical Ethics*, 159ff.

In the case of Hugh Finn, for instance, Hugh's family of origin was the first to oppose his wife's plan to disconnect feeding tubes. Then the Governor of Virginia took up the issue, as did a Virginia legislator. Their interventions profoundly deepened the tragedy in this case and complicated the severe burden carried by Hugh's wife and children. Such actions by politicians were ethically problematic in the extreme. The courts rightly rejected their petitions to reverse the wife's decision to carry out her husband's wishes. False hopes generated considerable social and legislative turmoil and thus prolonged the tragedy of Hugh's dying.

Ideology, Theology, and High Expectations

Complicating the relation of hope to healthcare decisions is a multilayered complex of interrelated mythologies. A myth is a powerful assumption that affects behavior and attitudes. The myth may contribute to misguided actions based on false premises, but a myth also captures a truth about common human experience and religious insights. False hopes and actions happen when a grain of truth is twisted and expanded until it is regarded as an absolute obligation. Absolutizing a partial truth is a fantasy or heresy that contradicts theological insights and ethical commitments. At least five myths can be isolated that contribute to the creation of false hopes in the clinical context and thus greatly complicate the task of providing appropriate medical treatment while safeguarding important faith convictions.

Death as an Ultimate Evil. The widespread notion of death as an ultimate evil finds expression in religion as well as popular folklore and medicine. Physicians often interpret the commitment to "do no harm" as requiring the prevention of death by any means necessary. The reason given for such warring against the inevitable is that death is thought to be an "evil."

A further and more far-reaching corollary is that medical science has a moral mandate to eradicate death. The latter is seldom expressed but seems everywhere promoted. The ultimate medical breakthrough would be the elimination of death.[18] And the technology to do so is being sought through research in the laboratories of chemists, geneticists, and engineers.[19]

There is no Christian mandate for science to eliminate death, nor is there much reason to believe that it will, but if death were to be eliminated,

---

[18] See, for instance, A. Harrington, *The Immortalist* (New York: Random House, 1969), and J. Kurtzman and P. Gordon, *No More Dying* (Austin TX: Learning Concepts, 1976).

[19] See ch. 7 on the artificial heart and cyborgs.

humanity would be most miserable. The success of science would be the undoing of quality human life and earthly existence as we know it, and a denial of our created being in God. If the grief of death is difficult to bear, think of the burden to the world if there were no death. What if Earth were now populated with all the people that were ever born?

The impact upon resources would be a horror too great to bear. Death has its rightful place in nature's scheme of things, as every religion attests. Personally painful as it may be, death is a created good; it is part of the cycle of life. Medical science and religion will do well to team their resources to dispel the myth of death as an ultimate evil, which is neither a biblical nor a commonsense perspective. Solid ethical and philosophical grounds exist for accepting death as having a necessary and rightful place in the cycle and symbiotic relationship of an ecological existence.

The Sanctity of Life. Most often associated with the right-to-life movement but found implicitly or explicitly in the justifications given for aggressive but non-beneficial treatment is another powerful myth often expressed in the ICU. If it is not death that is feared, it is "life" that is revered. The Missouri courts framed their refusal to allow the withdrawal of nutrition/hydration from Nancy Cruzan in terms of the "duty to preserve life." The reason given was that the liberty interests of the patient were subordinate to the state's general interest in "the preservation of life." As the Missouri court said, "The state's interest in life embraces two separate concerns: an interest in the prolongation of the life of the individual patient and an interest in the sanctity of life itself."[20] The horrifying possibility was that Nancy would be kept "alive" for the next three decades simply by administering nutrition and hydration. The Missouri court showed no awareness of the multiple ways in which "life" was being denied by the singular and simplistic focus on vital signs.

This reductionist notion of life is an abstraction from the meaningful or humanized living of the person. Some people settle for so little in seeking so much. The person is reduced to the vital signs, and worship is engaged at the altar of a biological idolatry. A biblical faith is considerably different and more profound. One's life does not consist in the tenacity or presence of biological functions, to paraphrase Jesus, but in the capacity and will to enter relationships responsibly and rightly to serve and enjoy God, the Creator and Redeemer. The Apostle Paul spoke of meaningful life in terms of his ability to serve the living Christ: "for me to live is Christ" (Phil 1:21). Life is a gift with which to live the abundant life. The Greek language

[20] *Cruzan v. Harmon*, 760 S W 2d, 408, 419 (1988).

distinguished *bios* from *zoe*—biology from the fulfilling, abundant life. That distinction is in sad need of rediscovery and implementation. To reaffirm something like a biblical understanding of personal life would help to dispel the false hopes that permeate the ICU.

Technological Wizardry. Jacques Ellul has decried the human reliance on technique and technology to the diminishment of our own humanity.[21] The world of high-tech medicine is a case in point. When the fear of death is wed to the notion of the sanctity of life, technology presents itself as the savior of humanity. Every human ill can be resolved at the altar of technological wizardry, and humanity is fulfilled by the mastery of technique. So people die attached to machines, waiting for the god of technological mastery to show its omnipotence and work another miracle.

Technology has its place and has been rightly and strongly affirmed by theologians and philosophers as one way of dealing with preventable tragedies and premature death,[22] but that valuable service has been absolutized and distorted to such an extent that it approaches a type of idolatry. Technology has come to master the mind that made it and demand loyalties to an extent usually reserved only for God. The ethical task for all medical enterprises is to maintain mastery over the technology employed for the sake of the human good.

Capitalism and the Financial Imperative. Few things motivate Americans quite so much as the desire for wealth or the fear of "socialism." Such hopes are cloaked in the rhetoric of the great American dream or the hope for the accumulation of wealth far beyond personal needs. The free market system is touted as the vehicle and tool for human entrepreneurial skills and economic salvation. The goal of riches provides justification for most any legal or quasilegal method for obtaining wealth. Even medicine has become afflicted with the ethical maladies of greed and materialism.

Medicine was once thought of as a noble profession, based on a calling to a compassionate and beneficent service, but it is now often perceived as the quickest route to riches and social prominence. Beginning medical students are more often motivated by the thought of wealth and prestige than that of serving people with healthcare needs. The entire medical system seems infected with the virus of the professional entrepreneur. Medicine has become big business, dominated by people motivated by profits rather than

---

[21] Jacques Ellul, *To Will and To Do*, trans. C. Edward Hopkin (Philadelphia: Pilgrim Press, 1969) esp. ch. 11.

[22] Cf. Callahan, *Setting Limits*, 148.

committed to patient care. To the degree that capitalism wins the affections of those in the system, medicine will lose its soul to the Faustian bargain.

Economic interests may also play a part in the resistance of physicians to accommodate the dying process. Technologies imposed interminably on the dying are sources of revenue, and numerous technologies may be applied. Disconnecting the machine or employing less exotic technologies reduces the revenue. One wit decried the advent of diagnostic related groupings (DRGs) as meaning "Da Revenue's Gone!" What clever insight into one of the dynamics of medicine as marketplace. Our metaphors and myths are powerful determinants of human action.

Miraculous Interventions. Expectations for a miracle is a final powerful dynamic leading to false hopes in the ICU. There is a widespread fascination about angels and miracles on the current scene. I share that religious tradition to a certain degree, but I have seen it so perverted that critical questions about the place and function of miracle seem appropriate and necessary. Bad religion peddles a perverse piety instead of truth and presents faith as belief in the incredible and unreasonable. The medieval jingle, "the more preposterous the belief, the more pious to believe," seems in full vogue in contemporary America.

A bogus and incredible hope often pervades the atmosphere of the ICU. It provides strength and resilience to those who wait on the dying as well as those who minister to their ailments. The downside is that the maximalist care[23] insisted upon escalates costs, denies the inevitable, and dehumanizes the dying. Such false hope insists on aggressive treatment even for people like William Bartling, a man dying of five different lethal diseases, including an aneurism on the brain and lung cancer. Likewise, the parents of thirteen-year-old Teresa Hamilton succeeded in forcing the hospital to allow them to take her home, convinced that she was only resting and would enjoy a full recovery.[24] Teresa had been pronounced brain dead in the hospital. She was a severe diabetic who showed a complete absence of neural activity and blood flow to the brain from successive brain scans. The hope behind maximalist treatment would require that all medical technology at our disposal be used to treat those overwhelmed by AIDS, or an anencephalic with severe respiratory distress.[25] False hopes are wed to the

---

[23] See George Annas, "Three Paradigms of Health Care," *NEJM* (March 16, 1995): 744–47, where he outlines the Maximalist/ Military Model; the Market Model; and the Ecological Model.

[24] *Hospital Ethics* (May/June 1994): 3.

[25] Ibid., 2.

technological mandate in a desperate but futile "hope" for a miracle. As Stanley Hauerwas puts it, such an expectation is neither realistic nor helpful.[26]

Theological and ethical reflection is needed to provide a more solid grounding for Christian hope and its correlative desires and aspirations. Christians do not hope for a few more days on *terra firma*, but for a life in the eternity of God the Creator-Redeemer. Miracles of curing and resuscitation are important, but they should not displace the hope of living in God's eternity. That hope has great importance in the ICU, assuring that people do not strive or die as those who have no faith. They know that death is not only inevitable, but finally necessary and desirable.

There is a point in the dying process that even aggressive interventions cannot succeed in restoring strength and health. The brain dead are and can be pronounced dead, according to current standards of medicine and law. We can say confidently that there will be no miracle cure no matter how fervent our prayers or wonderful the technology when the brain has died. Christians should not legitimate turning the ICU into a place in which we hope against hope, or engage in a fingers-crossed fantasy of wishful thinking. That journey into never-never land, where there is no death and loved ones go on forever, is neither biblical nor ethical. Denial has succeeded when hope springs unrealistically eternal and sophisticated technology is employed in the name of hope, or rationalized as being "medically indicated" or "medically appropriate."[27]

To be sure, hope has its place and is a necessary ingredient in patient recovery. Amazing recoveries should be given opportunity by providing an ample window for testing the possibilities, but beginning a treatment does not require a resolve never to quit. False hopes get refracted through the lens of denial and become the central model by which all desperate cases are treated.

This interpretation of the mythology of hope and its relation to "miracle medicine" is supported by a survey of physicians as to their attitudes toward futile treatment. Twenty percent of all physicians in three major medical centers in Texas indicated that the threshold of futility is zero percent. In other words, futility would obtain only if truly a zero percent

---

[26] Stanley Hauerwas, *Naming the Silences: God, Medicine, and the Problem of Suffering* (Grand Rapids, MI: Eerdmans Publishing Co., 1990).

[27] Jeffrey Swanson and S. Van McCrary, "Doing All They Can: Physicians Who Deny Medical Futility," *The Journal of Law, Medicine & Ethics* 22/4 (Winter 1994): 318ff.

chance of success for intervention exists—not one in a thousand or one in a million, but only zero percent. They argue that the belief in medical futility contradicts the principle of beneficence and/or non-maleficence. That attitude is hardly "scientific," but is a combination of questionable religious beliefs, defensive medicine, a denial of death, submission to the technological mandate, and personal styles in medical care.[28]

No neonate would be born damaged enough, no brain would be sufficiently dead, no body would be sufficiently overwhelmed by disease or massive injury to call it quits, to declare further treatment futile and senseless. Physicians would labor on, hoping against hope and selling hope to those who wait. Death has its place, but is often feared and fought to the last ounce of energy in the ICU. The role of the physician blends at times imperceptibly with that of the shaman and magician.

To be sure, the concept of futility, like that of interpreting the medical data, is problematic. Medical indicators may overlap with social values,[29] but futility is also based on careful assessments of particular medical conditions *vis à vis* the possibility for recovery of function and health. Certain metabolic or physical reactions might be restored even if health cannot be, of course. Such a "thin" definition of therapy leaves the patient in a state of limbo between life and death, heaven and hell, where medicine cannot cure but will not let go.

A Kentuckian was caught in that limbo. He was a victim of the futility versus "wait for a miracle" debate. The fifty-four-year-old went into cardiac arrest during an asthma attack. He suffered severe, irreversible, and extensive brain damage. The attending physician recommended that the ventilator be withdrawn and that a DNR order be placed on the man's chart. Nutrition/hydration would be continued. The hospital ethics committee supported the decision. The patient was then transferred to another attending physician who also supported removing the life support. The state cabinet for families and children, the man's legal guardian since 1991, then sought permission from the court to withdraw life support based on the recommendation of two attending physicians.

But the decision was challenged by the *guardian ad litem*, an attorney appointed by the court to represent the patient's interests, who argued that the state has no authority to substitute its judgment for the patient since there was no statement of his wishes. The attorney feared, he said, that

---

[28] Ibid., 318–26.

[29] Ron Cranford and Lawrence Gostin, "Futility: A Concept in Search of a Definition," *Law, Medicine & Health Care* 20/4 (Winter 1992).

withdrawal of treatment would "create the presumption under the law that everyone has to die and you eliminate any kind of miracle recovery in the future."[30]

One would have thought that the fact that "everyone must die" is by now a foregone conclusion. Death has a very high batting average, and in Christian theology, every person is destined to die. The human problem is to decide when it is morally and medically appropriate to battle for an extension of vitality so as to return the patient to optimal function, and when such actions manifest both bad faith and false hope. The interjection of "hope for a miracle" when the best medical diagnosis says the brain has deteriorated to the point of no return is hardly either good medicine, good morals, or good public policy.

Conclusions

The phenomenon of false hopes surrounding patients who are dying or whose illness cannot be reversed is a disturbing but entirely predictable and perhaps inevitable part of the clinical setting. Hoping for a miracle when medical science has done its best and the patient is overwhelmed by disease deepens the economic stresses on the healthcare system and prolongs the burden for families and other personnel. Such hopes are a combination of many factors, ranging from denial to sincere but problematic religious ideation. Having medical specialists disagree among themselves exacerbates the problem.

Medicine thrives on the very hopes it sometimes finds so troubling, however. Hope may provide energy and vision for productive enterprises, or it may consume precious resources. While hope steels us from the devastating effects of despair, it also makes us vulnerable to the wasteful and unnecessary. The paradox of living with hopes that are both true and false seems unavoidable. The dilemma belongs to life as *homo sapiens futurus*.

Both science and religion are challenged to confront and correct bad faith and false hope. The task is daunting and will never finally be completed, but the problem can be addressed and correctives attempted. Ministers and physicians, through their contacts with families in times of grief and decision-making, will discover "teachable moments," when the difference between hopes that help and heal and those that hinder and hurt can be discerned. In those moments the demands of faith and the rigors of reason become vital partners. Facing death requires a response of faithful

[30] Quoted in *The Courier-Journal*, 1 April 1996, B3.

obedience to the God in whose image people have been created. A Christian approach to clinical ethics will require that we not lapse into an idolatry of mortality or the false hopes of wishful thinking.

Chapter 3

HEALTH CARE AND THE FUTURE: FAITH PERSPECTIVES AND HEALTHCARE REFORM

> Of all the forms of inequality, injustice in health care is the most shocking and inhumane.
>
> —Martin Luther King Jr.

> America needs a comprehensive, understandable, affordable, national health insurance program with everybody in and nobody out.... National Health insurance—it's not just good health policy; it is a moral imperative.
>
> —Johnathon Ross, MD, Physicians for a National Health Program

Health care is an arena in which the interests of religion, science, and politics come together. Most religions place high value on human health and relate their understanding of God to the social and moral obligations to provide care for people young and old. Medical science is a combination of disciplines that attempts to educate the public as to the causes of disease and the primary modes of prevention as well as provide treatment modalities for those who suffer from injury or disease. Politics has been called "the art of the possible," where public resources are involved in the distribution of healthcare services to society at large.[1] Each is involved in current discussions and congressional efforts to plan a viable healthcare system for the future. The shape of that system and the scope of its services are of vital interest to all Americans.

---

[1] Richard D. Lamm, "Doctors Have Patients, Governors Have Citizens," in *Narrative Matters: The Power of the Personal Essay in Health Policy*, ed. F. Mullan, E. Ficklen, and K. Rubin (Baltimore: The Johns Hopkins University Press, 2007) 37.

This chapter examines the question of just how faith perspectives might contribute to public policy regarding health care. Without claiming that any single proposal is derived from Scripture, but seeking a system that serves the health interests of all Americans, this chapter explores moral action guides that might give direction for the future of health care in America

Health Care in America: Facing the Crisis

As "baby boomers" reach retirement age, increasing pressures are being placed on an already overstressed and, in many ways, failing healthcare system. Just how the needs of older Americans are going to be met is unclear. Predictions that Medicare will be bankrupt by the year 2020 are scary to most Americans but especially to citizens who are moving to new levels of dependency.

Affordability and accessibility are the two central issues in healthcare reform. Costs can be financially destructive. People on fixed incomes routinely have to decide between whether to buy prescription medicines or food and other life essentials. Most health insurance coverage is now through an employer. Without insurance, Americans are facing tough questions about paying for and thus accessing healthcare coverage. Even when people have coverage, the costs can be ruinous.

A sixty-two-year-old bivocational minister was recuperating from a fall from a roof he was attempting to repair. He fractured his skull and broke both wrists. He had no medical insurance and faced a bill of $98,000 for hospital care and $11,500 for an emergency helicopter flight to the hospital. Bills from the surgeon and physicians would follow. He had no idea how he would pay the bills. After "talking to God about it," he said, he discerned that he was to be patient. He is now living with a disability check from Social Security.[2]

C. L. has Gaucher disease, a disorder that affects internal organs and causes brittle bones. Gaucher previously has been an "orphan disease," meaning so few people have it that pharmaceuticals would not invest in research and development seeking a cure. Now, thanks to incentives from Congress, Genzyme has developed Cerezyme, from which it enjoys a gross profit margin of over ninety percent. But C. L.'s annual costs amount to

---

[2] *The Western Recorder* (Middletown KY), 17 February 2004, 1, 3.

over $600,000 a year for the medication alone. The costs are covered by insurance.[3]

Such are the ironies and disparities, if not inconsistencies and contradictions, in American health care. One person has curable injuries but is unable to pay the medical bills; another person runs up enormous costs for an incurable disease, but insurance pays the bills. Costs are not the only concern, of course, but certain questions are unavoidable. Just how much is a human life worth, for instance, and is there a way to translate that moral valuation into how much health care each is entitled to receive?

Several facts about the American approach to health care help extend and deepen an awareness of the problems confronted. For instance, nearly half of all bankruptcies in America stem from medical expenses, according to a Harvard study. Amazingly, these are people who have health insurance. Such bankruptcies affect two million Americans a year, counting debtors and their dependents. Of those filing for bankruptcy, three quarters had insurance at the start of their illness; thirty-eight percent lost their coverage by the time of bankruptcy when the illness led to loss of a job that provided insurance. The study pointed out that most Americans live just one serious illness away from bankruptcy. [4]

In fact, nominal care is so financially burdensome that many Americans do not go to a doctor. They cannot afford the deductibles or co-pay for expensive medications. Dr. Benjamin Spock, once called America's baby doctor, was reduced to begging for help from the public when he could no longer pay his medical bills.[5] Appeals for help with medical bills are often seen in conspicuous places like billboards and convenience stores.

Long-term or specialty care also remains a major problem. Senator Jay Rockefeller's mother died of Alzheimer's disease, which he called the worst and most humiliating of all ways to die—and one of the most expensive. His mother had required three nurses for round-the-clock care.[6] As he pointed out, most American families simply cannot afford to provide that type of care for an elderly parent.

The United States remains the only industrialized country in the world without a universal health insurance program. Canada, Britain, France,

---

[3] "A Biotech Drug Extends a Life, But at What Price?" *Wall Street Journal*, W, November 16, 2005.

[4] "Illnesses spur half of bankruptcies," *The Courier-Journal*, 2 February 2005, D3.

[5] "Wife of Dr. Spock asks friends to help with his medical bills," *The Courier-Journal*, 28 February 1998, A5.

[6] James Gannon in *The Jackson* (TN) *Sun*, 28 April 1994, opinion.

Germany, Japan, Sweden, Switzerland, and South Africa all provide an all-inclusive program.

America spends more per capita than any other country on healthcare costs, which continue to climb at double-digit rates. Health care is the third largest industry in the country, just behind agriculture and construction. In 2005, healthcare costs were $1.94 trillion, or nearly sixteen percent of gross domestic product (GDP). By 2014, that number is expected to reach eighteen percent, or $3.6 trillion and $11,046 per capita. We spend between forty and fifty percent more than the next most expensive nation.

Fifteen percent of Americans, or 46.6 million people, have no health insurance. Meanwhile, at both the federal and state levels, cutbacks are being made that most directly affect the least advantaged. Those who are elderly or have mental health problems seem most likely to have their social support system removed or severely restricted. A ninety-seven-year-old woman lost her Medicaid benefits even though she had congestive heart failure and was prone to falling. Another ninety-three-year-old woman was forced to leave a nursing home. She had both legs amputated, a heart condition, and an intestinal disorder.[7]

The problem is exacerbated by a fierce anti-tax movement that insists no state or federal taxes be levied to accommodate the needs of the dependent. At risk are those with infirmity, the emotionally or mentally disabled, and children with special needs, such as the autistic. Kentucky feared it faced a $450 million deficit and resorted to draconian measures toward those who depend on public assistance. Officials argued that patients could care for themselves at home! Funding the war in Iraq and responding to natural disasters in Louisiana has led to cuts in other services. Those "cuts" are to social services designed for the most vulnerable.

America can hardly pretend to be moral in its care for the poor, elderly, or disabled while such disparities in support and tragedies in personal suffering continue.

Healthcare Reform

Healthcare reform has been on the American agenda for decades. President Richard Nixon sought reform, as did President Jimmy Carter, who made a national health insurance program a major part of his presidential campaign. Health care was the top priority among voters in the 1992 presidential

[7] "Elderly and infirm fear loss of care as state trims rolls," *The Courier-Journal*, 20 July 2003, A1.

campaign between Bill Clinton and George H. W. Bush. Both political parties included healthcare reform in their platforms, but a concerted effort to overhaul the system failed during the first two years of the Clinton administration. A similar effort to effect change in the healthcare system has not been attempted—or even proposed—since that time.

The need for reform is apparently more widely recognized than is the political will for meaningful change. Former US Surgeon General C. Everett Koop called the American healthcare system "broken," and argued that "Band-aids" would not be adequate to cure its problems.[8] For years, healthcare costs far outdistanced the rate of inflation for every other sector in the economy. At the rate of inflation when Koop addressed the issue, healthcare costs would consume 100 percent of the gross national product by 2062.[9]

Factors in Escalating Costs

Healthcare costs have been driven by a variety of factors. Ironically, each cost factor is part of the image America projects of the finest healthcare system in the world.[10] The proliferation of specialties, for instance, attracts more physicians into exotic areas that, in turn, drives fees upward and reduces the number of family physicians. Specialists typically charge three times as much as the family physician for their services and are paid five times as much for a hospital visit.

A surplus of surgeons contributes to cost increases in a variety of ways. One estimate is that there are at least two million unnecessary operations per year, including circumcision and the placement of ear tubes. At one time, estimates were that fifty percent of all Caesarean sections, twenty-seven percent of hysterectomies, fourteen percent of heart bypass surgeries, and fourteen percent of back surgeries were not medically necessary. One surgeon joked that doing hysterectomies was his meal ticket through residency. The Institute of Medicine (IOM) estimates that nearly 100,000

---

[8] C. Everett Koop and Timothy Johnson, *Let's Talk: An Honest Conversation on Critical Issues* (San Francisco: HarperCollins, 1992) 85ff.

[9] Ibid., 86.

[10] R. LeBow, *Healthcare Meltdown: Confronting the Myths and Fixing Our Failing System* (Chambersburg, PA: Hood Publishing, 2003), cites this and other "myths" that hinder reform.

people die unnecessarily each year from complications experienced from unnecessary surgeries and medical mistakes.[11]

Another source of expense is exotic treatment that is medically non-beneficial. The story of the Lakeberg conjoined twins who were separated in Philadelphia is a case in point. Physicians in Chicago, where the girls were born, refused to do the surgery because of the severity of their condition. One of the children was certain to die from the operation; the other had a less than one percent chance of survival. But Children's Hospital agreed to the surgery at costs that exceeded $1 million. Indiana Medicaid then agreed to pay $600,000 for post-surgical care. Kidney dialysis was paid from public funds.[12] One twin died in surgery, as predicted. The survivor died three weeks before her first birthday. No family was with her. The commonsense budgeting decisions made by most American families seems to get lost in the high-tech, high-demand world of exotic, but terribly expensive, medical treatments even if they are of no or limited medical benefit.

The most powerful factor driving the astronomical costs of health care in America, however, is ideological, namely, free market capitalism. The hold over the American mind of private enterprise and free market forces is as powerful as it is ethically problematic. Reform efforts during the Clinton administration were undermined by a dishonest advertising campaign funded by the health insurance and pharmaceutical industries. Canada was maligned as providing less-than-expert care and marketing drugs that could not be trusted.[13] The accusations were untrue and self-serving, but effective. Americans have little conceptual capacity for thinking in terms other than a profit-based and driven system, and major corporations, who stand to benefit the most, do not hesitate to exploit the American sentiment.

At every point in the delivery of health care, individuals and groups seem primarily interested in how much profit can be made. The problem affects physicians, hospital administrators, pharmaceuticals, biotechnology groups, and the health insurance industry. Medicine has become infected with the malady of materialism, which represents a fundamental shift in the ethos that once shaped the medical profession. Mergers and acquisitions among the insurance giants have meant the end of competition among

---

[11] E. M. Wachter and K. G. Shojania, *Internal Bleeding: The Truth Behind America's Terrifying Epidemic of Medical Mistakes* (NY: Rugged Land, 2004) 57, citing the Institute of Medicine 1999 report, *To Err is Human: Building a Safer Health System.*

[12] See *The Courier-Journal*, 10 June 1994, B7.

[13] See LeBow, *Healthcare Meltdown*, 147ff.

private health insurers. The Anthem and WellPoint $18.4 billion merger resulted in the nation's largest insurer with $27.1 billion in assets. It lavishes $265 million in bonuses on a few top executives.[14] A few insurers now control most state and metropolitan markets.

Public policy in America protects pharmaceutical profits, both by shielding them from competitive bidding and ensuring huge profit margins. Comparison studies show that Americans pay a great deal more for brand name drugs than do the British, for instance. In the United States, Elavil costs nearly six times as much, Inderal nearly nine times as much, nordette seventeen times as much, and valium ten times as much. Captopril costs a patient $5,500.00 per year in its generic form. By the name brand, Capoten, it costs $122.95 for 100 fifty-milligram tablets.[15] Patents on drugs last for seventeen years, during which time pharmaceuticals zealously protect their monopoly and the government protects them from having to negotiate prices, as in the Medicare Plan D program.

Just how far the pharmaceuticals will go to protect high profits was shown in a report that Knoll Pharmaceutical Company sued to halt the publication of a pharmacist's study showing that a much cheaper drug was actually just as effective as its profitable Synthroid.[16] Knoll enjoyed near monopoly sales, eight-five percent, on the $600 million retail market for its synthetic hormone. The practice of delaying "undesired results" from studies contributes to the rising spiral in healthcare costs. Little wonder that prescription drugs are the single cost item most out of control.

Genentech recently offered what sounded like an ethical justification for charging so much for a new drug that might benefit patients with lung or breast cancer. The drug is Avastin, which will cost $100,000 for one year's treatment, doubling the current cost of the drug. The company claimed that the price must be weighed against "the inherent value of these life-sustaining technologies." [17] Leonard Fleck calls this statement "obviously self-serving disingenuousness. [18]

---

[14] Figures from the *PNHP Newsletter* (Summer 2005) 9.

[15] Figures taken from *The Courier-Journal*, 3 February 2004, A5.

[16] "Firm suppressed report showing cheaper drugs' effectiveness," *The Courier-Journal*, 16 April 1997, A3.

[17] See A. Berenson, "A Cancer Drug Shows Promise, at a Price That Many Can't Pay," *The New York Times*, 15 February 2006.

[18] Leonard Fleck, "The Costs of Caring: Who Pays? Who Profits? Who Panders?" *HCR* (May/June, 2006): 13. See also Dan W. Brock, "How Much Is One Life Worth?" *HCR* (May/June, 2006): 17ff.

Risky lifestyle choices are another factor in the escalation of healthcare costs. Americans have high expectations for health care that is effective, available, and inexpensive, while insisting on their right to a lifestyle that places health and life at risk. A study by the American Medical Association said unhealthy habits and violence cost the nation more than $42.9 billion in direct medical expenses. If related expenses, such as lost productivity, are added, the total is boosted to more than $189.1 billion annually. Smoking and alcohol abuse constitute the biggest drain, although violence and illicit drugs get more attention. Smoking costs the nation $22 billion, alcohol abuse $13.5 billion, and violent injuries add another $5.3 billion.[19]

Obesity is a major health problem in America, affecting sixty-one percent of all adults and fourteen percent of adolescents twelve to nineteen years of age. Heart disease, cancer, diabetes, stroke, arthritis, breathing problems, and psychological disorders are connected to obesity, which added nearly $117 billion to healthcare costs in 2000.

All-terrain vehicles (ATVs) are also popular—and dangerous. Kentucky and West Virginia lead the nation in deaths from ATV accidents. The death rate grew faster in Kentucky than any other state in 2002, according to the Consumer Product Safety Commission. Over five-and-a-half million ATVs are in use, resulting in 1,571 deaths and 136,000 injuries annually. Farmers and ranchers use them for work; children and adults use them for recreation. The speed, power, and weight of the four-wheelers make them popular in spite of the dangers. Thrill-seekers engaged in high-risk actions add to the burden of healthcare costs and to the social burden of caring for those who wind up with major injuries to the neck or spine.

The American healthcare system is thus a multifaceted problem that will require a multidimensional response. No single factor can be isolated as *the* explanation for the crisis in American health care. Those who plan for the future will need to design an approach that addresses many needs. C. Everett Koop called for a public-private alliance to provide a workable and accessible healthcare system. Beyond that, he said, there are three goals basic to reform: reasonable health care at affordable costs, the promotion of preventive programs, and an integration of the various levels of health care.

The Advent and Aims of Managed Care

One approach to cost control has been the development of Managed Care Organizations (MCOs) or Health Maintenance Organizations (HMOs).

[19] See *JAMA* (February 23, 1993).

These are typically for-profit corporations given a congressional mandate to control costs by imposing business practices in the world of health care.

The fee-for-service approach had built-in incentives for physicians to provide extensive diagnostic tests and referrals that escalated costs. George Annas called it the era of the maximalist/military model that regarded disease and death as the enemies and threw all available manpower and technology into the effort to defeat the enemy.[20] Economic considerations were beside the point; the victory over disease or injury was worth whatever it cost.

The market model now dominates health care, says Annas, which employs a number of strategies to control costs, such as limiting the number of tests and services. The tough-minded business approach makes those difficult calls that increase efficiency and thus maximize profits. Medicine is seen as a profit-driven business, the patient is seen as a consumer, and health care as a commodity governed by the laws of supply and demand in an open market. "Downsizing" and "hostile takeovers" of hospitals and other healthcare provider services became commonplace. Free market forces, not the central management of government or the even looser approach of physician fee for service, were to be the savior of the American healthcare crisis.

Columbia HCA, with its aggressive billing strategies and cutbacks in medical services, became a symbol of market techniques applied to health care. High-powered business executives were making decisions affecting the health of patients. The ethical criticism was that the wrong people were making the wrong decisions for all the wrong reasons. One physician-turned-hospital-administrator filed a class-action lawsuit against an MCO, claiming that they were "killing our patients." When profits are more in focus than patient need, the system itself is morally bankrupt. Even choices between tests or treatments may be based on cost-saving strategies rather than efficacy or risks to patient. However, the cheaper drug may be chosen in spite of its having a far higher likelihood of adverse reactions, including cardiac performance, renal function, depression of the central nervous system, pain at site of injection, flushing, nausea, and vomiting. A great deal more is involved in moral medicine than cost considerations. As one writer put it, managed care is "a wrong-headed philosophy of health care."[21]

---

[20] George Annas, "Three Paradigms of Health Care," *NEJM* (March 16, 1995): 744–47.

[21] See J. P. Kassirer, "Editorial," *NEJM* (July 6, 1995): 51.

Such judgments also carry over into allocation questions. Who benefits, and by what standards should costs versus benefits be decided? The question cannot be avoided since some procedures have extremely low success rates and very high price tags, such as bone marrow transplants for certain types of advanced cancer. For that reason, some plans restrict or deny coverage. The public reaction to such decisions was not always in touch with the medical reasons for such denials. When Humana denied payment for a complete hysterectomy for a woman who had been treated for cervical cancer, it was following standard medical practice. Specialists from across the country agreed that she did not need the hysterectomy. Even so, her tearful appeal gained the sympathy of the jury.

Profits and benefits for the insurance company or shareholders are not good reasons for medical cutbacks. But the best medical practices do not always follow from the most expensive procedure possible, as in the case of the conectomy. The best medical intervention is based on a calculation of the benefit or burden to the patient. Other areas restricted by HMOs were more strongly disputed, and rightly so. Cutbacks in hospital stays for childbirth, for instance, were successfully attacked. Both mother and child need more careful and extensive monitoring than one day in the hospital following childbirth. The reaction of the public and Congress was predictable and strong. Some important changes in HMO rules are already in process, but the goals of efficiency, preventive medicine, standardized treatments, and affordable insurance are necessary features of the stewardship required in health care. As Annas puts it, what is needed is an ecological/stewardship approach that takes seriously the need to preserve scarce resources, control costs, and maximize access.[22]

Malinowski calls for a new era in which physicians "internalize cost-conscious calculations" and demonstrate a social conscience. "Rationing health care resources is necessary," he says. At a minimum, there should be "a presumption against trying expensive treatments which are highly unlikely to work." The eras of medical ethics in which there was a "no-concern-for costs mentality" are over, he says.[23]

The larger economic question is how to pay for healthcare services in an efficient and cost-conscious fashion. The ethical question is closely related: how can we fairly and equitably share the benefits of medicine to

[22] Annas, "Three Paradigms," 746.

[23] M. J. Malinowski, "Capitation, Advances in Medical Technology, and the Advent of a New Era in Medical Ethics," *American Journal of Law & Medicine* 22/2–3 (1996): 339.

persons of equal standing as citizens? Among the problems to be faced is how to assure fairness in the allocation of benefits so as to maximize preventive medicine without denying care *in extremis*. At present, most patients with advanced cancer receive bone marrow transplants, which offer very little chance of health improvement, while many women cannot obtain prenatal care and twenty-five percent of children are without healthcare coverage during their first years of life.[24] The enormous amounts of money spent do not always yield maximum dividends.

Universal Health Care

Proposals for universal health care suggest an altogether different approach to healthcare financing. Advocates say their approach is affordable, efficient, and equitable—all the features that seem desirable or necessary in a morally supportable healthcare system. David Jones, founder and former CEO of Humana, says everyone should get the same healthcare coverage available to senators and representatives. Why, he asks, should our elected officials get care that is not available to the rest of us? That comment is rather remarkable since it comes from a man who played a major role in the move toward for-profit health care in America. He is now a convert to the notion of universal care.

Physicians for a National Health Program (PNHP) support a single-payer system, believing it would save nearly $200 billion a year in total healthcare costs.[25] Once implemented, the physicians believe the annual healthcare costs for families could be reduced to affordable levels from the average of approximately $10,000 that is now paid. Private insurers would be prohibited from selling coverage that duplicates services or procedures, but could sell additional benefits not covered under United States Health Insurance (USNHI), such as cosmetic surgery. The physicians' plan would set reimbursement rates for physicians and other healthcare providers. Drug prices would be negotiated with pharmaceuticals; they would no longer be able to set their own prices arbitrarily.

The plan also calls for a conversion to a nonprofit system over fifteen years. Financing would be by the sale of US treasury bonds. There would be no payments for loss of business profits, only for real estate and equipment. The system would be paid for by a payroll tax on all employers of 3.3

---

[24] Ibid., 335.

[25] G. D. Schiff et al. "A Better-Quality Alternative Single-Payer National Health System Reform," *JAMA* 272/10 (September 14, 1994): 803ff.

percent, a five percent tax on the top five percent of income earners, and a small tax on stock and bond transfers. Corporate tax loopholes would be closed; the Bush tax cut would be repealed.

Needless to say, the industries that benefit from the current system will strongly resist such proposals. The physicians' group is taking direct aim at the enormous profits made by people who know little about medicine but a lot about business strategies for profit making. CarePlus CEO Mike Fernadez was paid $330 million from the merger of two Medicare HMOs in South Florida. He had bought into the business two years earlier for $38 million. Other CEOs make millions in salaries and bonuses and brag that they are realizing the great American dream, but it hardly seems moral to build financial empires at the expense of people who go without basic medical care.

Health as Belonging

The drive for healthcare reform is rooted in a basic moral concern for human well-being. Health is not just another consumer good that can be sold to the highest bidder. Having health and strength are basic to the pursuit of a person's vision of life and happiness. One must be able to avoid catastrophic illness and injury and ward off or bear pain and suffering in order to have sufficient physical and emotional energy to succeed at one's chosen profession. Without health one can hardly have equal opportunity or a personal standing in community, or what Wendell Berry calls "membership."[26]

The notion that health care is a "commodity" fails to capture the essential meaning of health to the person. Healing seeks wholeness—the restoration of the self from an inner alienation that goes to the core of one's existence. Disease is an ultimate dis-ease, for it threatens one's very existence, one's very being. The loss of health is like no other loss since it also portends a terrible turn in the road toward loss of life. A surgeon once described the horrible impact of hearing over the phone that he had been diagnosed as having diabetes. To the physician breaking the bad news, it was just another call about an ordinary and widespread illness. He had seen it all before until the disease and its symptoms had become commonplace, but it was not commonplace to the physician who is now a patient. To him, it was the end of one world with its goals and assumptions and a confrontation

---

[26] Wendell Berry, *Another Turn of the Crank* (Washington, DC: Counterpoint Books, 1995) 86.

with a new and terrible reality. Disease threatens one's well-being in every dimension.

To commodify health care is to objectify the person. Health care becomes something *like* a radiator, shoes, chair, or a piece of clothing. It can be bartered and sold, withheld or given. It becomes an economic transaction, but health care is *different* because it has to do with health and well-being. Loss of health is a threat to the person's capacity to pursue values and goals essential to life and its meanings.

Health Care as a Right or a Privilege?

The question of health care as a right or privilege can better be addressed once the issue of health is located as the heart of the issue. Should health care be available on the basis of the ability to pay or because it is a social good and those who need care are part of a community of sharing?

The language of rights pervades the vocabulary of American jurisprudence and ethics. A right is moral shorthand for the entitlements that belong to individuals or groups. At one level, of course, the pursuit of health is within the prerogative of each person. Each is free to exercise, avoid high-risk behavior and dangerous habits, eat the right foods, and secure sufficient rest for the body to rebuild its energies and ward off disease. That freedom is called a negative right—no one is obliged in any way to provide the benefit claimed, but the pressing question in America is whether there is a positive right to health care—a societal obligation to provide healthcare services to those who cannot afford it, and if so, on what terms?

The United Nations Declaration of Human Rights (1948) listed health care as a universal right, but the United States has never declared it a constitutional right. There is, however, an implicit recognition of a basic right to health care. Under the Emergency Medical Treatment and Active Labor Act (EMTALA), no person may be refused health care who presents to an emergency room, but that is a far cry from the type treatments that persons need in order to maintain optimum health. Medicaid and Medicare were created in order to provide healthcare coverage for groups thought particularly vulnerable—the poor, the disabled, and the elderly. Such provisions are a type of social safety net for the least advantaged.

The reimbursement rate for hospitals and physicians under Medicaid is so low, however, that many hospitals and physicians do their best not to accept Medicaid patients. Access to decent health care for this population depends upon the extraordinary devotion and selfless service of physicians and other healthcare providers. Typically, the poor are treated in substandard facilities and by inadequately trained personnel. The hospitals are plagued by lack of staff and outdated technology. A cancer unit's radiation unit at a Brooklyn hospital was dubbed "the killer" because it destroyed both cancerous and healthy tissue in nearly equal proportions. Insured patients across the street, on the other hand, enjoyed state-of-the-art technology.[27]

The pressures to survive in a heated market economy are making it difficult to provide even nominal levels of charity care. Physicians are among the first to feel the financial crunch. Their response has been to cut back on free treatments. A recent study shows a twenty percent decline in such care over the past decade (1995–2005). The Center for Studying Health System Change, a nonpartisan research group, says volunteer charity care seems to be disappearing. A nonprofit group that provided free checkups for needy children going to summer camp has closed its doors in Washington. According to the chief executive of the community health centers of south-central Texas, volunteer charity care is nonexistent.[28]

The market effect is also reaching the organizational level of charity care. LeBow notes a metamorphosis taking place in religious healthcare institutions. They began as "charity care" organizations, living out a mission of humanitarian service in the name of love and mercy. Their shift from nonprofit to for-profit, he says, represents a shift from a calling steeped in caring and compassion to what can only be called "a market-based entity where business goals rank first."[29]

Three Approaches to Justice

Whether there is a social obligation to provide a system for health care involves an intense debate among philosophers and theologians. The question is one of distributive justice. How are social and medical resources to be made available so that each person shares in the burdens and benefits of the healthcare system? At least three philosophical approaches to justice are prominent in current discussions: utilitarianism, egalitarianism, and

[27] See J. D. Arras and B. Steinbock, eds., *Ethical Issues in Modern Medicine*, 5th ed. (Guilford, CT: Mayfield, 1999) 618. See also T. Beauchamp and L. Walters, eds., "Justice in the Distribution of Health Care," in *Contemporary Issues in Bioethics*, 6th ed. (Belmont, CA: Wadsworth, 2003) 39ff.

[28] "Proportion of Doctors Giving Charity Care Declines," *Washington Post*, 23 March 2006, A09.

[29] Lebow, *Healthcare Meltdown*, 123.

libertarianism. Being aware of the basic assumptions of each may help to understand the profound differences that emerge among people of good faith and solid intelligence.

Utilitarian perspectives permeated the study by the President's Commission, which concluded that society has a moral obligation to provide for the health needs of its population. Even so, it stopped short of saying people have a right to health care. The commission also felt that a free market cannot be trusted to deal equitably with the distribution of benefits. A more intentional approach is needed that will keep an eye on costs while providing a decent level of health care.[30] The commission embraced a vision of the good society that promotes the health and well-being of its citizens. Morally, society must be committed to the relief of the burdens of suffering and premature death. The moral commitments of society are questionable if it neglects those who are ill.[31] Even so, cost-saving is not only morally responsible but necessary. The principle to follow is "access for all to an adequate level of care" without imposing excessive burdens.[32] No one in serious need should be left without care, and the care that is provided should meet the standards of sound medical practice.[33]

Egalitarian theory makes an even stronger argument for social obligations regarding health care. John Rawls argues there is a "social contract" to advance the good of all persons within society.[34] Rawls accepts the fact that people are never entirely equal either in their natural abilities or in their general level of health, but he argues that we should all work to minimize our inequalities and the economic and social distances between us.

Rawls begins with an intriguing exercise in moral imagination. He calls for people to place themselves in "the original position" in order to imagine the types of agreements that would be accepted as constituting a fair or just social contract. Everyone is both determined to protect his or her own interests, he says, and is rational, that is, motivated to reach agreements that are not counter-productive or self-defeating. In that hypothetical position, persons are behind a "veil of ignorance," deprived of knowing their situation in life as determined by such matters as race, gender, religion, or ethnic

[30] See "The President's Commission for the Study of Ethical Problems in Medicine and Biomedical and Behavioral Research," from *Securing Access to Health Care*, vol. 1 (Washington, DC: US Government Printing Office, 1983).

[31] Ibid.

[32] Ibid.

[33] Ibid.

[34] See John Rawls, *A Theory of Justice* (Cambridge, MA: Harvard University Press, 1971) 11. See also his *Political Liberalism* (New York: Columbia University Press, 1993).

origin. Contractual obligations require people to accept eventual outcomes and differences among themselves. Thus, they must imagine that they may be among the least advantaged as well as hope they will wind up among the most advantaged in the various strata of the society that emerges.

There are two principles, Rawls says, that would be accepted by rational, self-interested people. The first is that each person is to have an equal right to the most extensive basic liberty compatible with similar liberties for others. The second principle is that social and economic inequalities are to be arranged so that they are both reasonably thought to be to everyone's advantage and attached to positions and offices open to all. Rawls' brand of egalitarianism is not a social ironing board in which everyone is meted out exactly the same and no more. Differences are both tolerated and encouraged as long as the openness is to the benefit of everyone as a whole.[35]

No rational person would agree to an arrangement in which the least advantaged would simply be ignored or abandoned by those who benefit the most. The natural processes of genetic chance and social circumstance create great disadvantages for some and advantages for others, but behind the veil of ignorance, one does not know how social processes or personal circumstances will work out. Thus, no rational person would agree to a system that does not care for the least well off since they might wind up in that situation.

Libertarian thinkers strongly object to arguments about obligations and rights that the strong owe the weak or the wealthy owe the poor. The ideological grounds of libertarianism are found in Robert Nozick's book *Anarchy, State and Utopia.*[36] Rush Limbaugh is arguably libertarianism's most vocal media advocate. Nozick says there are entitlements for those who earn them or who deserve them based on their hard work in the free market system. There are actually very few absolute rights, he says, but among them is the right to private property. As long as property is acquired or transferred by just processes, one's title to that property is nearly absolute. Certainly property rights are not undercut by the needs of others, he says.

---

[35] See R. M. Veatch, "Justice, the Basic Social Contract, and Health Care," in *A Theory of Medical Ethics* (New York: Basic Books, 1981) 250–80.

[36] R. Nozick, "Distributive Justice," in *Anarchy, State and Utopia* (New York: Basic Books, 1974) 149–69.

H. Tristram Englehardt[37] follows Nozick in saying there is no basic human right to the delivery of health care, not even to a decent minimum. He says there is no moral or philosophical basis to justify coercing people to give up property in order to provide for the healthcare needs of others. Government cannot violate the (property) rights of some in order to accommodate the assumed rights of others. Redistributive economic arrangements are thus inherently unjust. Using the tax code to effect social goals such as saving lives or benefiting quality of life with advanced medical technologies is a social *choice*, he says, not a matter of social justice. The failure to provide health care may be stingy, uncharitable, and uncompassionate, but it is not unfair or unjust. It is certainly no violation of anyone's "rights."

A social Darwinism permeates such libertarian arguments. Differences among people, he says, come primarily from two sources. One is a natural lottery, the other a social lottery. The natural lottery,[38] he says, is primarily the workings of natural forces. The brutal vicissitudes of nature are unfortunate but not unjust, since they are "acts of God." Since no one (government) is directly to blame, no one can be charged with the responsibility for rectifying such ills.

The social lottery, says Englehardt, serves to separate people in terms of economic or social strata. Some people prosper as a result of love, self-denial, affection, and mutual interests, he says. They convey resources to one another, and these people develop fortunes. Those who lack such associations do not prosper. Their economic disadvantages are not from the malevolence or unjust actions of others. These people suffer from misfortune, but there is nothing unfair about social and economic differences.

Injury or illness caused by the actions of other people is a different matter. Government does have a responsibility to protect citizens from harm. Insofar as it fails to do so, the government should compensate the injured. To this extent, there is an obligation to provide health care supported by taxation. Beyond that, however, all that is needed is a system of free market healthcare insurance, which should be privately and voluntarily purchased by individuals or by group initiatives. A free market economy makes no attempt to achieve equality of health care despite the constant appeals for charitable support.

---

[37] H. Tristram Englehardt, "Rights to Health Care," in Beauchamp and Walters, *Contemporary Issues*, VI, 64–71.

[38] Ibid., 66.

Biblical Guidelines for Healthcare Reform

Discussions about justice as a major principle of medical ethics are an immediate point of contact with the Bible. Doing what justice requires is a central mandate of the biblical ethic. It stands as a corollary to God's demand for universal love, which informs and tempers the temptation to think of justice simply as the workings of any system of law. Civil law should seek to do justice, but it is never simply an embodiment of biblical justice, as prophets from Moses to Martin Luther King Jr. have reminded the people of God.

Ironically, religious groups tend to identify with charity or love but are less likely to be champions for justice, says LeBow.[39] Even so, the concern for justice as it relates to healthcare reform needs the light of the biblical witness. Scripture does not mandate any particular system of health care, but it provides profound insight into the nature of justice and its importance for human relationships.

Justice: A Call to Right Relations

Just how the teachings of the Bible might be related to public policy in a pluralistic society requires and deserves careful attention. Turning to the Bible is a matter of developing a perspective informed by the biblical materials, which provide a wealth of wisdom and moral guidance for those interested in and committed to shaping the moral community. All people stand to benefit as they become part of the stories of the people of God. We are all called to effect justice and to live out the righteousness of God in the social order.

The term "justice" translates words that occur over a thousand times in Scripture. Stories of the Sabbath and Jubilee years, the prophetic judgments upon clever but unethical business practices, the stories of deliverance and conquest, of the wise but worldly king Solomon deciding between two women which of them should have the child each contends is her own, and Jesus' parables of the righteousness of the kingdom all are woven into this richly textured and multifaceted mandate for God's people in social relations.

The biblical writers did not deal with justice as an abstract, formal principle. No philosophical discourses are to be found to resolve the debates that surround us. The biblical writers focus on concrete personal and

---

[39] LeBow, *Healthcare Meltdown*, 125.

corporate relations and the moral requirements of living in the context of the covenant community. There is no language of rights, as such, but the concept is everywhere apparent.

The laws of hospitality formed the core and essence of Hebrew social morality,[40] constituting something like a social contract between the wealthy and powerful regarding their obligations toward the poor and powerless. The reasons given for protecting the poor and landless were not abstract and philosophical but historical and personal. Leviticus 19:33–34 is a pivotal passage: "When a stranger sojourns with you in your land, you shall not do him wrong. The stranger who sojourns with you shall be to you as the native among you, and you shall love him as yourself; for you were strangers in the land of Egypt: I am the LORD your God."

Exodus 22:21 is also a blunt statement of the mandate and its rationale: "Do not oppress the stranger or sojourner; for you were strangers in Egypt" (cf. Lev 19:33; 25). The moral justification for the generous treatment to the sojourner turns on a historical reason: at one time they "were strangers in the land of Egypt." They survived because of the mercy and generosity of strangers. The moral mandate for treating strangers with toleration, decency, and compassion thus amounted to a type of "do unto others as you would have them do unto you," as the reminder about their life in Egypt shows.

Strong sanctions were applied to Israelites who refused to live by the laws of hospitality. The divine censure for disobedience to this central moral mandate was at issue in the story of Sodom and Gomorrah (Genesis 19:5–25). According to the story, the messengers of God came to visit with Lot, who was a resident alien in this foreign city. Lot, in terms of contemporary images, had his visa, but the messengers had not registered with city officials. Upon hearing that their laws had been violated, the men of Sodom came out to humiliate and injure Lot's guests. Such inhumane treatment violated the law of hospitality. The stranger was to be treated with dignity and respect, not humiliated and injured by harmful and despicable acts.[41]

When God's own people violated the laws of hospitality, the requirements of justice turned on them. A parallel account to the story of Sodom in Genesis 19 is found in Judges 19:22ff. This story portrays the Benjaminites of the city of Gibeah as guilty of inhospitality against an old

---

[40] Thomas Ogletree, *Hospitality to the Stranger: Dimensions of Moral Understanding* (Minneapolis: Fortress Press, 1985) 1.

[41] See Ezekiel 16:48–50; Matthew 11:23.

man and his concubine. The men of Gibeah wanted to injure the man but settled for taking their wrath out on his concubine. She was ravaged and killed. The people of the remaining tribes of Israel made war on Benjamin for this degrading treatment of fellow human beings. Both the Sodom-Gomorrah and the Gibeah stories capture the divine mandate of providing care and sustenance to strangers and sojourners and to take care not to abuse or injure them.

The point need not be belabored by examining all the passages that form the so-called "laws of hospitality." The claim of human rights never appears. The construct of human rights so basic to contemporary thought did not emerge for centuries after the Bible was written, but the concept is implicit in the story of the people of God and their obligations under the laws of hospitality. Strangers and sojourners, in effect, had "rights" to toleration and access to human resources that were protected by covenant law and appeals to the moral responsibility of the community of faith.

The eighth-century prophets were strong champions of the justice of God and its demands upon the human community. The poor and vulnerable were regarded as special objects of God's love and protection, a note that also permeates the New Testament teachings. Jesus inveighed against those who fashioned laws so they might take economic advantage of "widows and orphans" (Matt 23:14; Mark 12:40; Luke 20:47).

Jesus' Miracles of Healing

Jesus' miracles of healing also provide support for a Christian approach to health care. Jesus did not denigrate the body in favor of a transcendent spirituality, nor did he offer people in pain the assurances of blessed relief in the hereafter. He demonstrated concrete cures in the name of God's love. He healed people who were blind, epileptic, hemorrhaging, paralyzed, or comatose. Lazarus was resuscitated (John 11:43), an action that now takes place almost routinely in hospitals and elsewhere. Jesus healed people to enable them to glorify God in a body that was whole and with a mind that could will to serve and obey God (John 9:4). Paul spoke of the person as a psychosomapneumatic (mind, body, spirit) whole (1 Thess 5:23)[42] who is the object of God's concern.

[42] The passage from Thessalonians is a prayer that "your spirit, soul and body may be kept sound and blameless," which seems to outline the dimensions of human personhood.

Miracles of healing also show that disease and disability are not God's way of punishing people for misdeeds. Religious explanations of suffering based on punishment theory or a notion that God is attempting to teach a lesson otherwise to be missed were emphatically rejected by Jesus (John 9:1–4), as they had been in the story of Job. Disease is the experience of the absurd in natural processes, human history, and family life. God is not present as the one inflicting the pain and suffering, but as the Comforter, attempting to bring healing or to provide wisdom, strength, and courage.

Jesus' dealings with people with certain illnesses also captured the special pathos of the ostracism often meted out upon those with disfiguring or mysterious disease. Those with leprosy, epilepsy, menstrual problems, and mental illness were ostracized and regarded with contempt. Instead of shunning such people, Jesus reached out and touched them, bringing both healing and acceptance into community (Matt 8:3; cf. Lev 13:24; Num 5:2). Such stories have motivated and informed compassion for those with AIDS and other dread diseases.

Stories involving Jesus' crossing the invisible lines of ethnicity that created barriers to community also inform the Christian attitude toward justice in health care. Whether it was a Gentile centurion or a woman with an issue of blood, Jesus reached out to people ostracized by prejudice and superstition. He welcomed them into the community of love and acceptance. Few illustrations so powerfully capture the truth of health as belonging. Jesus had to counter the effects of the social and religious stigmatization that accompanied disease and disability.

The all-too-common human reaction to such people is a combination of sympathy and revulsion. Such ambivalence contributes to apathy toward the illness or hostility toward public policy that might otherwise accept responsibility for health care. The people of God have the mission of encouraging compassion; they are to reach out and accept those rejected by bigotry, fear, or ignorance. Sympathy for the suffering is to be deepened, and revulsion is to be confronted as the lack of love and acceptance of persons that it is. Fear of the diseased is a sign of superstition and anxiety that prevents a moral response to those in need of health and care.

The story of the Good Samaritan (Luke 10:25–37) is a central paradigm of the demands of love and a just response to human health crises that transcends any ethnic or religious identity. The Samaritan happened upon a Jew who had been beaten, robbed, and left for dead. Unpredictably and surprisingly, the Samaritan took the wounded man to an inn some distance away and paid for the man's care from his own resources. Jesus approved his actions as an embodiment of the love for neighbor and justice in community that is required of all people. The story illustrates the relation of justice to love. The individual Samaritan becomes a paradigm or representative of corporate life in a community seeking the demands of social justice. Justice is love seeking the well-being of the neighbor through public policy.

Health Care and the Moral Community

A system of health care builds upon the altruism at the heart of the good and just society. Providing the needed good of health care has the cohesive effect of binding persons to one another, as Emerson said. Whether based on the utilitarian concern to reduce pain and increase happiness or the egalitarian principle of doing to others as we would have done to ourselves, a number of vital social values are affirmed.[43]

First, an inclusive healthcare system affirms social solidarity.[44] People are dependent upon one another. Libertarianism builds on the deception of the self-made, independent person. We all stand on others' shoulders and depend on others during times of special vulnerability.

A second value is that it builds on a vision of human flourishing in mutual dependency. People can do a great deal together that can never be accomplished by individual efforts. A cooperative effort to ameliorate pain, reduce suffering, and prevent premature death will provide depth to social existence.

The third value is that life itself is shared at very basic levels. Few things evoke such gratitude as a second chance at life. The excitement and pure joy among the Italians for the gift of donor organs from an American child killed by bandits is symbolic of the social contribution made by a system of generous care.

A comprehensive healthcare system will embody three truths that tend to be neglected in a competitive capitalist environment. One is that our relationships are not primarily or exclusively economic in nature. We are people in relationship, and human values must be cultivated if the social dimensions of our personality are to flourish. Second, health is not simply another commodity. The body is not just another object, like a vase or a

---

[43] See T. Murray, "Gifts of the Body and the Needs of Strangers," *HCR* (April 1987); and in Carole Levine, ed., *Taking Sides: Clashing Views on Controversial Bioethical Issues*, 7th ed. (Guilford, CT: Dushkin, 1997) 350–58. Murray is writing about organ donation but the points he makes are also germane to health care.

[44] Murray, "Gifts of the Body," in Levine, ed., *Taking Sides*, 353.

piece of household furniture. Health is a gift like no other, and its loss is a threat to human life at its most profound levels of meaning. Third, we are all involved in an ongoing quest to relieve suffering and pursue the common good. The society that knows the truth of the solidarity at the heart of our common humanity and cultivates ways to bear one another's burden is one that has a claim to being moral.

Justice and Healthcare Choices

An approach to healthcare reform that reflects, if it is not based on, the biblical concerns for social justice and righteousness in human affairs will include the following features.

First, we should openly declare a right to health care. We need not find the language of rights either in the Constitution or in Scripture to support the concept. The basic moral obligation to care for the poor and disenfranchised lays a strong foundation for regarding health care as a moral right—the obligation owed one another to a basic level of health care. We cannot deny health care to the poor and claim to be moral or consistent with the biblical message of justice and righteousness.

The fact that religious leaders can be found supporting a wide variety of approaches to justice reveals a loose commitment to biblical norms. Christians are not free to adopt just any theory of justice.

The free market approach to the distribution of health care seems a transparent rationalization for the selfish protection of economic privilege. George Soros argues that very few wealthy capitalists believe in the free market system. What is "free" to them is not free to others. He says capitalists live by government largesse, not by their own clever designs for making money.[45] United States public policy is not socially neutral. Government now favors those who have the most capital, ostensibly on the theory that they will create jobs for others. More realistically, it is wealthy barons in collusion with powerful politicians in a quid-pro-quo arrangement that keeps each in places of prestige and increasing wealth at the expense of those less favorably placed.

Second, a right to health care is a claim to basic humanitarian support in a healthcare setting. It is not a right to everything that might be desired or anything that might be wanted. Healthcare policy should reflect the commonsense approach to allocating the benefits of health care that affect

---

[45] George Soros, "The Capitalist Threat," *The Atlantic Monthly* (February 1997): 45–58.

positively the vast majority of the population. No one is required to break the bank in order to provide for health care. We can neither break the public's back in providing health care nor neglect the basic needs of the poor and claim to be moral.

There is a certain truth in Englehardt's distinction between misfortune and injustice that should be noted. While he takes the issue too far as a way to rationalize harshness toward the unfortunate, he nonetheless points to a problem that is of concern. There are some misfortunes that should be accepted and not become the occasion for doing all that is technologically possible in attempting to correct the uncorrectable. Calculating costs must also include the wave of special care obligations that follow such persons as they move through the social system. The amount of continuous care required through a lifetime is simply staggering. There is no moral obligation to engage in aggressive, but non-beneficial, efforts to keep the terribly damaged infant alive. Aggressive treatments are morally required when there is a reasonable ability to restore or assure optimal functioning. Interventions that result in a life with major deficits are questionable.[46]

Finally, concerns for justice allow and require at least some degree of utilitarian calculus in order to decide allocation decisions and healthcare priorities. The state of Oregon has taken a bold step in this direction. Their Medicaid plan assures access to health care for the poor. But they use calculations of what healthcare procedures produce the best outcomes for the money invested. Bacterial pneumonia is treatable and curable, and the antibiotic is inexpensive, so it is covered. Treating anencephalic newborns, on the other hand, is not covered. Vast amounts of money can be spent in valiant efforts to prolong the life of an anencephalic neonate, but nothing can be done to assure anything like a normal life or an open future. These infants are born without a brain or only a partial brain, and nothing can be done to correct that problem. They do not receive aggressive care in Oregon; they are allowed to die a natural death.

Oregon also employed a utilitarian calculus across the spectrum of healthcare options. A list of 709 procedures was drawn up. All procedures up to number 589 were covered; those above that number were not. Mastectomy for breast cancer is covered, but reconstructive surgery is not. The latter is regarded as cosmetic and is deemed medically unnecessary, even if desirable.

The biblical notion of wisdom and the stewardship of resources inform the Christian acceptance of utilitarian calculus. Instead of having vague

---

[46] See ch. 5 on the right to die.

notions of what is "right" or "wrong" or being led by the emotions of the moment toward those who are diseased, damaged, or injured, utilitarian thinking introduces commonsense logic and reasoned calculations. The problem of what to cover is at least in part subject to quantification.

The biblical concern for wisdom is close to utilitarian thinking. There is nothing unethical or un-Christian as such about utilitarianism. Whether used to deal with ways to maximize happiness and diminish pain or as a way of calculating the best way to use scarce resources, it requires using the mind, an exercise that is both necessary and moral. God gives the ability to reason; we need all the help we can get not to allow emotions to rule when innocent suffering is at stake.

The biblical imperatives for justice and love finally force the issue of which approach to health care should be supported and pursued as public policy. In this regard, the proposals for universal health care are most persuasive. No other approach recognizes the centrality of the mandate to provide care. We need not claim the Bible mandates a particular plan among others, but it does mandate that any plan include all people on equal terms.

The proposal that best fits the demands of justice and love in social relationships seems to be that of a single-payer or universal healthcare coverage. Physicians have it right in supporting a national health plan, and it is unjust for legislators to provide the best health care for themselves while refusing a similar arrangement for all Americans. No patient should be reduced to begging because of healthcare costs, nor should any patient with treatable wounds or illnesses be turned away for reasons rooted in costs and financial incentives. The most obvious way to institutionalize our care for persons is to remove the financial barriers to health care. That move is past due; it should be done and done now.

Chapter 4

AGING AS AN ASSAULT ON HUMAN DIGNITY: SPIRITUALITY AND END-OF-LIFE DECISION-MAKING

Cocoon, a science-fiction movie about aging and the assault on human dignity, deals with human responses to diminishing powers and the reality of death. The main characters are four couples (really four men) in a retirement community in Florida. Their lives reflect not only the circumstances of middle-class affluence, but the inevitable tensions created by their place in the later stages of life. Their routines are designed to maintain the three pillars of human dignity—mental agility, physical ability, and financial security.

In the retirement home, patients are wheeled around in chairs, carried around on beds, or hobble around with canes or crutches, while the lucky, healthy ones walk around with reasonable strength and mobility. The enjoyers, the survivors, and the casualties can all be found in the community.[1] Death lurks there, too. One patient is being resuscitated; all the others wince, knowing they also live in the valley of the shadow of death.

The lead characters are the walking ones, hopeful in the midst of all the signs of deterioration of body and mind of which they are constantly reminded. Memories clutter the pathways of their minds—blessing them while haunting them. Sexual powers are waning. One man returns a flirtatious invitation by saying he would send a man "if he finds one." Their eyes are dimming—cars are parked and covered when driver's licenses are lost. Their bodies are losing strength. Youthful dancers taunt "the old guys" who now know they cannot compete on the dance floor—or most anywhere else. They are typical of an aging population. They embody the hopes and fears, the dreams and pains, of all of us in our older years.

---

[1] M. Lewis, *Aging in America: Trials and Triumphs* (Westport CT: Americana Healthcare Corporation, n.d.) 17–18.

Dignity and the Hope for Health

Three of the men enjoy swimming in the pool of a mansion being leased by strangers who have placed large stones from the sea into the pool. Then they notice the changes; they are being rejuvenated. Muscles are restored, eyesight radically improves, sexual strength returns. Joy and hope return. Joe is cured of cancer—miraculously, says his doctor. Their partners are brought to the pool of health and hope. They all discover energy and strength to spare. They may be old, but their bodies have the vigor and vitality of youth. Their energy and enthusiasm for life seem boundless. Others notice the change and wonder why a few seem to be growing younger and their powers restored while the rest continue on the course of slow but noticeable decline.

Bernie is the odd man out among the four. He notices the rejuvenation in his friends but refuses the invitation to swim in what he dubs "the fountain of youth." He shares the belief that human dignity requires accepting the limits of mortal life established by nature. Like Charles Lindbergh[2] and Dr. Francis Moore,[3] Bernie believes people should accept the approach of death with equanimity, not undue anxiety. For him, the mad scramble to extend the vital signs a few more days is both unnatural and futile. "Nature dealt us our hands of cards and we played them," Bernie says. "Now at the end of the game you want to reshuffle them."

Ironically, in a heated argument with Joe, it is Bernie that reveals the secret to a crowded room of elderly residents. They all stampede *en masse* to the pool, which by now stands nearly full of the huge rocks placed by the new tenants. They run, walk, dive, jump into the pool, eagerly hoping to experience the vigor others had recently recovered.

What they did not know was that the pool was prepared by Antereans (extraterrestrials) for their friends who had been buried in Atlantis for centuries. The Atlantans had survived in "cocoons" that must now be incubated to bring them to life, but the hopeful elders banged the rocks around trying to open them—their curiosity won out over caution. Enthusiasm for benefits caused them not to care about costs or risks. Only the return of the Antereans stopped the destructive trespass, but by then it

---

[2] Charles Lindbergh, famous aviator and inventor, retired to Maui and prepared for death when diagnosed with cancer.

[3] F. D. Moore, "Prolonging Life, Permitting Life to End," *Harvard Magazine* (July/August 1995): 46–51 (ch. 32 of his book *A Miracle and a Privilege: Recounting a Half-century of Surgical Advance* [Washington, DC: Joseph Henry Press, 1995]).

was too late—the damage had been done. The Antereans in the "cocoons" were dying before they could be brought to life. As Walter, the lead Anterean, put it, "All this time to energize the pool. Now the life force is completely drained, and I'm not going to be able to bring them back."

Aging and Anxiety

This sci-fi venture imaginatively captures the tension and anxieties within us as we confront aging. The constant awareness of waning powers, the nagging reminders of the ravages to mind and body created by the wear and tear of our years in space and time, and the grim prospects of death all belong to this common venture of life. We are faced by the four dread horsemen of dependency, disability, dementia, and death.

We may be encouraged by the poet's invitation: "Come and grow old along with me / The best is yet to be / The last of life for which the first was made."[4] The romantics we have with us always. Optimistic and glowing portrayals of old age are not uncommon. They reflect the hope captured in *Cocoon*—that medical science is on the verge of miraculous breakthroughs that will assure us all of perpetual health and happiness. It is a portrait of life without debilitation and death, but we all know better.

The Problem of Human Dignity

Aging is an assault on human dignity. Science may delay the end result and slow the process, but it cannot and will not eliminate the inevitable. Our powers are under attack, and our dignity is at stake, but human dignity is a difficult term to define. Part of the problem is that dignity is not a thing, like a part of the body or brain that can be identified as an organ and dissected for its component parts. Dignity is of widespread interest and concern among ethicists and theologians, but there is no consensus as to its meaning.

A root of the term refers to honor or rank, and thus connotes having the status of a dignitary. Dignity, by this account, would refer to a sense of decorum or air of presenting oneself as when it is said that he or she "walks with dignity," meaning that one has a certain air, or style, that conveys pride and a sense of self-worth, but that is hardly sufficient, since dignity would be possessed by only a few.

---

[4] Robert Browning, "Rabbi ben Ezra," in D. H. S. Nicholson and A. H. E. Lee, eds., *The Oxford Book of English Mystical Verse* (Oxford: Oxford University Press, 1917) 106.

The Enlightenment emphasized a more universal meaning, referring to dignity as an inherent worth of a person as a human being. *Dignitas* was almost synonymous with *humanitas*. To be a person was to have dignity because one belonged to the human family and was endowed with the powers of reason.

A more distinctively religious meaning uses the term to connote certain inalienable rights that are not to be denied since they are given by God. The Vatican document *Respect for Human Life* says that dignity consists in "being endowed with a spiritual soul, (given) moral responsibility and a calling to beatific communion with God."[5] The Vatican says everyone has dignity from the moment of conception to the last moment of life.

A fourth meaning is phenomenological. Factors to be considered include such matters as self-esteem, self-identity, and the entitlements necessary to support and protect the self as person. The elements of self-identify and a sense of well-being in society and the world become the central considerations. By this approach, dignity is not simply a "given," except in the most abstract sense. Dignity must be related to a sense of self that provides the person with sufficient ego strength and self-regard that life is not only meaningful but is defined by profoundly important personal values.

When approached in this way, there are four pillars of self-esteem or personal dignity: ground-of-meaning beliefs; physical health and bodily control; mental agility, including comprehension, integration, and communication; and economic security, sufficient to provide security from poverty and the self-destructive angst that deprives one of the essence of personhood in a social context.

People are unique in their capacity to reflect upon the processes of events and circumstances in which they are involved. They are constantly aware of what affects them for good or ill and the forces that move them toward a future that can be anticipated and, to some extent, modified. Like Ben and Joe in *Cocoon*, people are risk-takers, seeking ways to increase their joys in the midst of failure and frailty. Only *homo sapiens* can reflect upon the past, contemplate the present, and anticipate the future. Only people can alter circumstances and thus change outcomes. In this is human dignity.

---

[5] See, for instance, Congregation for the Doctrine of the Faith, *Instruction on Respect for Human Life in Its Origin and on the Dignity of Procreation: Replies to Certain Questions of the Day* (Rome: The Vatican, 1988) 2. See also Pope John Paul II, *The Gospel of Life* (New York: Times Press, 1995).

Human reflective and volitional abilities make possible the experience of dignity—an evaluation of the self as unique, reflecting the powers of God. People are, at least to a considerable degree, autonomous, as Kant argued so well. Autonomy implies agentry, the ability to reason and reflect upon possibilities and alternatives toward the future. People are not simply mindless flotsam blown by the winds of change and circumstance. We become persons—a self—in the awareness of the uniqueness of being human. The dignity of being human is a gift from God.

Autonomy also involves mobility or action. Persons are defined by movement, by altering circumstance, or by shifting location. Action is the ability to pursue values, beliefs, and intentions, and thus achieve the goals and purposes that make life meaningful. Mental agility is thus linked to physical vitality. The powers of the brain marshal the strength of the body for actions reflecting thought and mind that define our being in the world, with others, and before God.

Growing older both contributes to our sense of dignity and threatens to destroy it. Like the retirees in *Cocoon*, we are made wiser by the distilled wisdom of our years, but we also feel the inevitable erosion of the powers by which our sense of self has been defined and out of which our dignity has been discovered. Little wonder that dementia, whether senility or Alzheimer's, has become a part of our conscious dread of the future. Few deteriorations are feared so much as the loss of conscious mind when the body remains biologically tenacious. There is no dignity in this death of the self.

The Bible and Aging

The Bible reflects and acknowledges this paradox at the heart of the human situation. The writer of Proverbs says the hoary head is a crown of glory (Prov 16:31). The aged were said to deserve respect out of the fear of God (Lev 19:32). The giants of the faith were people who attained stature, power, and prestige in the fullness of years. The vitality of Moses between the age of 80 and 120 and the longevity of Methuselah were celebrated and regarded as the rewards of divine blessing. Such stories embodied the human hope for length of years and strength for living.

The biblical writers suffered no illusions. Old age takes its toll. Isaac (Gen 27), Jacob (Gen 48), and Eli (1 Sam 3:2) all experienced blindness, Eli in his ninety-eighth year of life. And the pathos of the aged and infirm King David was told with care and compassion. Old and weak, his body cold and unresponsive, the young and beautiful Abishag was brought to comfort him, but he remained cold and unmoved. There are volumes in the writer's comment that Abishag "cherished the king and ministered to him, but he knew her not" (1 Kgs 1:4). The Teacher warned young people to enjoy God while strong and vigorous because "the evil days" would come in which "you will say 'I have no pleasure in them'" (Eccl 12:1). The twelfth chapter of Ecclesiastes could be called "A Lamentation on the Prospect of Aging, Dependency, and Death."

Beneath our anxiety about lost powers is a gnawing fear of dependency, that we shall finally be placed totally at the mercy of those who have power over us, that we are losing our autonomy. We are all vulnerable, of course, but vulnerability increases with age and declining powers. Jesus once reminded Simon Peter that "when you are young, you dress yourself and go where you will; but when you are old someone else will dress you and carry you where you do not want to go" (John 21:18).

One of the dynamics behind prejudice and discrimination against the aged is the fact that "the aged remind the middle aged of their own imminent destiny," as May points out.[6] We loathe what we fear. Our place of residence, the disposition of our property, the types of medical treatment to which we shall be subjected, or the persons with whom we associate may all be determined by someone else—the younger, the stronger, the more independent. Our life may come under the control of a stranger who is either ignorant of or unsympathetic to our values, our hopes, and our dreams—the very things that constitute our sense of self. As the Turnages say, "Perhaps the most miserable, the ultimate role change is to dependence!"[7]

Diminished independence is a threat to our dignity. The fears of being ignored, taken for granted, isolated, or abandoned compound the anxiety attached to diminished powers. The psalmist expressed this fear for all of us: "Do not cast me off in the time of old age; forsake me not when my strength is spent" (Ps 71:9).

A related but distinguishable fear is that of becoming a burden to those who love and care for us. Family systems can be terribly disrupted by the need to care for the dependent elderly. Professional or vocational schedules and objectives may be jeopardized by the commitments of time and energy required when acute illness or chronic disease takes its toll. Modest estates intended for children or grandchildren are also placed in jeopardy by the

---

[6] W. F. May, "Who Cares for the Elderly?" *HCR* (December 1982): 33.

[7] Mac and Anne Turnage, *Graceful Aging: Biblical Perspectives* (Atlanta: Presbyterian Office on Aging, 1984) 37.

staggering costs of health care. Some elderly do not worry about abandonment—they know only too well of the support systems provided by loving families. No expense or sacrifice is too great to provide for the needs of the older, dependent mother, father, or grandparent. It is precisely this concern for the well-being of the larger family that distresses many elderly as they anticipate declining health. Elderly suicides may be as much prompted by love for others as by despair over dependency, as the van Dusens[8] showed. For some people, death is not so much to be feared as the damaging sacrifices others may make on our behalf.

Myths and the Dignity of Persons

Our approaches to aging reflect our perceptions of reality—our understanding of what is true and basic and thus valuable and to be preserved. We live by myths and metaphors, by shared assumptions and images of truth. A myth is a vision or perception of reality out of which we structure our patterns of behavior both personally and socially. The "myth" may be true or false or a combination of the two, but it is a powerful determinant of attitudes and behavior. Ageism, like sexism and racism, reflects the myths by which we live. Life and human dignity, death and human destiny, for instance, are vital categories of belief in our approaches to aging. The basic ingredients in our essential worth or what constitutes the dignity of being human are components of the myths by which we relate to aging. Two primary myths are those of the self, portrayed in terms of economic worth and biological basics.

Economic Reductionism. One of the most powerful myths dominating attitudes toward and treatments of the elderly is that of economic reductionism: the value of the person as person reduced to an economic calculation. Eric Fromm noted its devastating effect on ego identity. The person is taught to understand self-worth by the formula "I am = what I have and what I consume."[9] Financial factors are certainly important to human well-being. An economic calculus is both reductionistic and depersonalizing, however, resulting in patterns of discrimination against those whose productivity is measurably less than others. For years, women who were homemakers suffered from this pattern of discriminatory evaluation. Declining health, strength, and energy in older persons make

---

[8] Dr. and Mrs. Henry P. van Dusen made a pact to commit suicide, leaving a note expressing their faith in a prayer: "O Lamb of God that taketh away the sins of the world."

[9] Eric Fromm, *To Have or To Be?* (New York: Bantam, 1976) 15.

them the objects of ridicule and discrimination by those who measure value by the yardstick of financial productivity.

The vitality and health of President Ronald Reagan helped for a while to challenge the idolatry of youth in America. He showed that life after sixty-five (indeed after seventy-five) need not be a "vast wasteland" of inevitable decline, powerlessness, and lack of respect. As he approached the biblical promise of "three score years and ten," he attained the zenith of his career and the apex of power in the world. His robust health and vigorous lifestyle encouraged millions who faced the "graying of America."

Compare the optimistic mood regarding aging that now prevails to that reflected in a valedictory address given in 1905 by William Osler, dean of the Johns Hopkins University Medical School. The great work of the world was accomplished by people between the age of twenty-five and forty, he said, and men above sixty should retire since they were largely useless. He even referred to the novel by Trollope that depicted chloroforming people over the age of sixty-seven. A rash of suicides unfortunately followed, as did a harsh but humanitarian criticism of his views.

Osler's satire was missed by many of his hearers who turned angry and intolerant toward a great physician and what they felt were his views. He had regrettably reflected the myth of economic valuation and was thought to be making an assault on the dignity of older persons. Economic reductionism calculates the value of the person as person as a measurable and quantifiable factor—one's economic contribution.[10] Only as one works is one valuable—does one have dignity—according to the cruel calculus of utilitarian valuation.

Good morals require constant diligence against such reductionist schemes that figure so prominently in public policy. A person's life and worth consist of more than the job to which one is attached. The myth is still alive and does its tawdry, demoralizing damage to the ego strength of the retired. It creeps into institutional policies of hiring and firing and measures us all by the dollar sign.

Ironically, the advances in the war on ageism have been accompanied by a new mythology of older people as healthy, sexually active, engaged, productive, self-reliant, and financially independent. If Reagan helped to stem the tide of pessimism about aging, he also became a champion of the new myth. The myth he embraced was as dangerous as the image of the

---

[10] See further in the chapter on healthcare reform. State budget cutbacks typically start with the vulnerable, including the mentally ill, the aged, and the disabled.

elderly as decrepit and dependent. The older population is neither all as healthy as Reagan, nor as unhealthy as William Bartling, the seventy-six-year-old dying of five lethal diseases and wanting to be allowed to die. The new mythology tends to undermine political and moral support for the sick and needy by ignoring or masking the existence of illness and poverty. How else could Reagan seriously argue that there were no hungry Americans, or, if there were, it is because they are ignorant of available help?

The new myth of aging also places burdensome and impossible expectations on the elderly. Those who are not "healthy, wealthy, and wise" are regarded as having only themselves to blame. Those who are advancing in years are expected to diet and exercise with Jane Fonda or Richard Simmons so as to preserve the trimness and energy of youth. What is obvious is that the new myth is no more tolerant or respectful of old age than was the old. The "slings and arrows of outrageous fortune" simply cannot be excluded from the experience of advanced years, as Reagan learned after leaving office. Alzheimer's is no respecter of persons. Decay, dependency, and death are here to stay in spite of our successes, fantasies, mythologies, and unrealistic expectations.

Biological Reductionism. One of the great contributions of Christianity to public health has been to support the notion of the worth and dignity of every person. The biblical insight is that human dignity rests upon our creation in the image of God (*imago dei*) and our being the objects of God's love and care in incarnation and redemption. God's love for each is the ultimate basis for our respect for one another. Our dignity is not our own; it is from God. It stems not from nature, nor from civil law, but from the Creator. Human dignity rests not on measurable worth, but on our uniqueness as persons. Our sense of self-worth reflects a value granted by God.

Care must be taken that respect for persons does not succumb to the pagan fiction of biological idolatry, however. The dignity of persons is a far more complex and cherished reality than the presence of vital life signs. One's life consists of more than pulmonary and respiratory function—important as these are to our being a person. The life of dignity and worth is biography as well as biology. It is a world of sense and feeling, of passion and play, of recognition and being recognized, of social interchange and spiritual introspection. It is the world comprised of a sense of self—the consciousness made possible by the awareness that we *are* and that we are in pursuit of a worthy and open future. That sense of self is a non-negotiable minimum that lays claims upon the community for respect and prohibits violations of personal space and religious values. Our self is in and before God. Dignity consists in knowing this if I know little or nothing else at all. My own struggle with hopes, fears, and dreams is the self of dignity that I cherish and call others not to injure or abuse. To have one's sense of self attacked by those who have the advantage of power but do not respect our values or religious beliefs is to be dehumanized or depersonalized.

Dignity under Assault: Two Images

Two images of powerlessness, vulnerability, and frustration illustrate the assaults on human dignity confronting the aging. The first is that of a comatose minister—another of those tragic stories on the way to the Patient Self-Determination Act (PSDA). He had served as a pastor in Kentucky for a half century. Struck down by a massive stroke, he was taken by a county EMS crew from a single automobile accident to a hospital. The trauma unit specialists succeeded in resuscitating him, though he had been unconscious and not breathing for at least twenty-two minutes. He was placed on a ventilator, diagnosed as having massive brain damage with a prognosis of persistent vegetative state.

Life support systems were continued even after it became known that this man had duly signed and executed a living will, specifying he did not want aggressive treatments that would only make his dying a more prolonged process when there was no reasonable hope of recovery. The trauma team had no way of knowing that, of course, and thus applied all the technical wizardry at their disposal in restoring signs of life to a man who was dead. A loving daughter and distraught wife insisted on discontinuing the ventilator based on long conversations and their intimate knowledge of how strongly he felt about this matter, but the hospital team persisted, supported by an intimidated and defensive administration and a fearful, conservative hospital attorney.

During a month of being medically manipulated contrary to his will, the minister was weaned from the ventilator. He was then transferred to another hospital. Once breathing on his own, but in a "persistent vegetative state," he was placed in skilled nursing-home care. The modest estate he intended for family then became exhausted by hospital and nursing-home expenses.

The tragedy in this portrait of human suffering is multifaceted. Deprived of strength of mind and body by an acute trauma, this man of dignity was deprived of any semblance of human being-ness by the persistent denials of his last will and testament. In the name of "restoring life," the will of the strong was imposed on the unconscious and dependent patient. Such imperialism in the exercise of power by medical professionals lays waste to the moral basis for medical treatment. It amounted to no less than an assault on the man's dignity—the integrity of his body and mind and the respect due his religious and moral convictions.

Such are the fears with which we all live. The prospect of such manipulations belongs to any health crisis, but it is especially acute for the elderly who walk ever deeper into the shadows of disability and death. The medical systems designed and developed to help us have become potentially an enemy assaulting the last semblance of human dignity.

The second image of elderly vulnerability is that of the distraught wife in this case. Physically strong, she had suffered no loss of vitality from acute or chronic illness. Her helplessness was an impotence imposed by the medical system and its ability to deny decisions to family. She felt strongly that her husband's wishes should be honored, but discovered only passive resistance cleverly disguised as sympathy from the hospital staff. "If I were in your place," said the attending physician, "I would want him disconnected, too." His sympathy was only seeming, however, since it was in his power to order the disconnect. He refused "on moral grounds," he said, since the man was breathing (howbeit, by the aid of a machine)—and his heart was beating (howbeit, because of the forced oxygenation).

The woman's pain was a complex mixture of many frustrations. The sudden loss of a beloved husband was grief enough. The pain of separation from him in the cloister of the ICU was added to the burden of watching him sustained without hope of recovery. The great contradiction was in loving him as person and seeing the way he was treated as a patient. He had been so objectified by hospital procedures that he was no longer recognizable as the person she knew so well. The machine contradicted his religious opinion and life commitment that whether he died or whether he lived, he was the Lord's (Rom 14:8). What mattered was not that vital signs be prolonged a day or two—or weeks that turn into terrible months—but that his dignity in dying be preserved.

Her repeated demands were politely but firmly denied under the guise of medical responsibility. They must, after all, "first, do no harm"! The frustration of powerlessness became a cruel mix with the grief of loss. An explosion was inevitable. Passionate anger expressed to the physicians was condescendingly regarded as "stress anxiety" brought on by fatigue and grief. Kentucky's archaic misogynist laws added to the burden. At that time it was not clear that she was the legal guardian for an incompetent spouse's medical treatment decisions, though he would have been for her. An ancient conspiracy of ideas now combined with unfortunate events to produce a tragic situation.

Her anger also turned inward. Exasperated and desperate, overwhelmed by bills and new responsibilities, she collapsed on the hospital floor. The emotional and psychic pressures were too much for her. The questions, problems, and deadlines overwhelmed her carrying capacity. Her own life was now in jeopardy. A modest estate and retirement income had already produced anxiety enough for each of them to bear. Their prudent lifestyle attested limited means as well as moral and religious commitments to frugality, industry, and simple living. They really believed that a person's life does not consist in the abundance of what is possessed (Luke 12:15), but in the value of persons and love and mercy. They were now subject to a system that did not fully share or appreciate those values. They were learning a harsh lesson about financial stability and economic accountability in the midst of sudden death and loss.

Having shown mercy, she now received none. The sympathy given by administrators, physicians, and nurses was undoubtedly genuine, but she felt it was rejecting, contemptuous, and punitive. The medical team accepted no responsibility for the events that threatened her life. He was a "prisoner in the ICU,"[11] held captive until the last farthing was paid. Her simple but all-important request was that her husband be allowed to die—that he not be maintained with what he had for years regarded as a pagan denial-of-death routine in the ICU.

In that denial there was an all-out assault on human dignity. Unwittingly, healthcare officials denied their own dignity by refusing the courage necessary to challenge a pagan idolatry of vitalism, by denying the patient's dignity to have wishes respected and not despised, and they denied the dignity of family (wife, daughter, and brother), all of whom joined in the request to cease and desist with death-prolonging measures. The Patient Self-Determination Act was long overdue.

Dignity, Dying, and Scarce Resources

From a faith perspective, dignity consists not only of an autonomy dependent on strength, but in the strength of acknowledged dependency.

[11] G. J. Annas, "Prisoner in the I.C.U.: The Tragedy of William Bartling," *HCR* (December 1984).

We are all dependent, for we are limited and finite creatures. Finitude is a fundamental threat to human existence, of course. Death is a near-ultimate threat, and our fear of death is another factor in our prejudice against the elderly. Being human, we know that we are to die, but that is also the source of our greatest anxieties. Woody Allen captured the human ambivalence: "I really like the idea of dying and going to heaven," he said. "I just do not want to be there when it happens." At another time he said, "I always wanted to be immortal, not by dying and going to heaven, but just by not dying."[12]

In *Cocoon*, Bernie rightly saw that human dignity is preserved most profoundly in our ability to face death with equanimity. There is no dignity in the scramble to exhaust vital resources needed by our children and theirs. Unprepared for death, we are desperate to live at any cost. We lay waste the future in the mad struggle to clutch more and more until there is less and less for anyone. Governor Lamm's sage advice is difficult to embrace: "We all have a duty to die," he said, "and get out of the way and make possible a decent future for our children."[13] The enormous drain on Medicare made by expensive health care is symbolic and telling. The vast percentage of all Medicare funds are spent on hospital and health care for the last two weeks of life. The treatments for the dying elderly are costly, but they are not medically beneficial.

The story line in *Cocoon* graphically portrays the problem. The masses of elderly desperate for "rejuvenation" may overwhelm the system. There is only so much life-strength to go around. There is no dignity in the mad and selfish scramble for resources that jeopardizes the health of the future. Community requires caring and planning for the well-being of all—the vulnerable infant as well as the vulnerable aged and those in between. Sentiment and fear can blind us to the moral necessity of calculating the requirements of justice. A blind egoism not only subverts justice, but makes personal survival tactics a morally suspect, "ends justify the means" mentality.

Even Bernie, the moral protester who embodied the concern for the future in *Cocoon*, finally broke under the weight of grief. He could accept the prospect of his own death with greater grace than he could accept the loss of his beloved wife, Rose. When she died, he panicked and, in desperation,

---

[12] Woody Allen, quoted online at "Words to Live and Die By," http://home.earthlink.net/~ggsurplus/words2.html.

[13] See Richard D. Lamm, "Health Care as Economic Cancer," *Dialysis and Transplantation* 16 (1987): 432–33.

took her to the pool where he pleaded with Walter for the life-giving powers: "Just one more. Help me; help me," he cried.

"I'm sorry," said Walter. "It's too late. The pool doesn't work anymore."

The pool is our human and technical capacities or our non-renewable resources developed over millennia. There is enough to sustain us all if we do not panic and insist on consumption beyond what is needed to provide sufficient sustenance for a decent life, but, enraged by fear and panicked by the fear of death, we may make life impossible for generations yet to come.

Little wonder that some sociologists are predicting a major intergenerational clash. Already, the "sandwich generation"—those caught with responsibilities for children as well as parents—is complaining about the stresses on expenses and the burdens of personal demands. We are moving toward an era in which thirty percent of all Americans are dependent—counting those over sixty-five and those under eighteen. How many dependents can a productive society carry? Medical costs are now nearly sixteen percent of the gross national product. More can certainly be borne and undoubtedly will be. The moral claim of the needy is strong, and resources are still available, but at what point will the resistance begin and the resources dry up? The near future may bring the realization that the pool has stopped working.

The political clout of the older populace is considerable. By some accounts, the AARP is the most influential lobbyist in Washington. Certainly it is a major force in shaping public policy. In 1900, only three million people in the United States were over sixty-five, accounting for four percent of the population. There are now thirty million (twelve percent). Estimates are that by the year 2030 the percentage will rise to between twenty and twenty-five percent. On a daily basis, 1,550 people turn sixty-five. The moral task is to minister to the infirmities of the dependent ill without jeopardizing the resources necessary to the health and well-being of future generations.

Human Dignity and the Crisis of Faith

One of the disturbing themes in *Cocoon* is its dismal perspective on the human prospect for the future. The outcome of the story is hopeful only in a sci-fi way—the promise of life and hope, of restoration of health and strength lies in being rescued by extraterrestrials. All the "hopeful" and mobile residents of the retirement village decide to escape to Anterea. The climax of the film is a secular ascension reminiscent of Elijah's being taken into heaven (2 Kgs 2:11). There is no death, just transition. As Ben tells his grandson, "Your grandmother and I are going away. When we get where we're going, we'll never be sick, and we'll not grow older, and we'll never die."

The scene is reminiscent of comforting stories told by the dying to their sorrowing survivors, or the promises made from pulpits about life after death and the grace and power of God that restores health and provides endless bliss. "There will be no more death," says the writer of the Apocalypse, "for the former things have passed away" (Rev 21:4).

The difference between the hope centered in technology and that centered in a future promised by God is profound. The Christian hope lies *beyond* death, not *instead* of death. The movie is a secular version of the human hope for immortality. One by one, the movie discounts the props of religious hopes. Science helps, but cannot cure; friends are important, but destined to die. The message is clear: hoping for immortality is wishful but futile thinking.

Cocoon is a challenge to the basic affirmation of religious faith regarding death and the afterlife. The dilemma of modern people confronting the apparatus of medical science is that they have no hope beyond this mortal coil. The desperate resort to ventilators to sustain the dead and the futile but expensive manipulations to which we subject our bodies seem testimonies to a hope without faith. Beneath it all—the physician's frantic efforts to prevent death by every means possible, the public's demand for biological immortality and its expectation, or the quick medical fix for every physical ailment, and our desperate drain on resources for the future—lies the persistent existential *angst*—the fear of death.

The movie is not anti-religious; it is just not a religious or Christian statement of faith. Even Bernie is portrayed as a victim of fate, not a person of faith. He is a Stoic, not a Christian. He does not hope to see Rose beyond the grave—he takes no risks in the freedom of grace. He is to be fertilizer for the earth—his final contribution to future generations. From the dust he has come; to dust he returns. His life has been a passive waiting; his facing death is a picture of pathos. He is left hopeless and helpless.

Toward a Spirituality of Aging and Death

The crisis confronting us in our aging and dying is a crisis of faith and of hope. We can be robbed of our dignity only if we are a people deprived of a firm belief that God's plan for us extends beyond this earthly existence. We can rightly encourage science to extend life expectancy, to cure where possible, and comfort always, but to preserve life signs beyond the possibility of restoring sentient existence is a worship of the life principle; it is biological idolatry, rooted in the fear of death.

Religion's greatest contribution may be precisely as a way to deal with death anxiety. All religions offer belief systems as a way of coping with the threat of death. The fear of death may well be the fundamental cause of the origins and continuity of all religions, from the animistic tribalism of native cultures to the highly sophisticated organizations that now compete for believers on the world stage. Religion has a grasp on the human heart because it has tapped the fundamental fears of the human mind. Religion proposes to answer the great questions raised by those confronting the reality of their own demise.

Christianity and other major religious traditions deal openly with death anxiety. The fear of annihilation, for instance, is countered by the teaching that the self is moving toward completion in God. The Christian's hopes and dreams for wholeness and health in bodily existence are expressed openly in the Bible. The Apostle Paul spoke of a spiritual body (1 Cor 15:44) and the Apocalypse captured the hope for overcoming deformity and disease. In God's eternal city, John said, there would be no pain or tears, and no more death, since death will have been swallowed up in life (Rev 21:4). Families facing the grief and loss of death often gain comfort from the Apostle Paul's statement that "if for this life (*bios*) only we have hope, we are of all people most to be pitied" (1 Cor 15:19).

Religious faith is a powerful coping mechanism for those facing death and a solid source for guidance as one prepares for death. When the marvels of medicine reach their limits, religion reminds the believer that death is not an ultimate threat. Salvation is in God's promised future, not in the wonders of medical science or the wishful thinking of rescue from outer space. The human hope is ultimately in God and the hope that delivers us from the bondage of fearing death (Heb 2:15). We need not fear death, for we believe in the God "who has brought us to life again unto a lively hope" (1 Pet 1:3).

Cocoon appeals to the hopeful imagination that looks forward to a type of technological panacea, or the promise of immortality from scientific technology. The Antereans (ETs) had already conquered aging and death, according to the story line. They never die; there is no illness, no Alzheimer's, no disability. *Cocoon* is an idyllic projection of human hopes in a fascinating and entertaining story of science fiction. Beyond its entertainment value, the greatest comfort it can offer is that science is moving toward a more perfect future for dealing with the diseases that afflict humanity. For the present, people must deal with the harsh realities of disease and the experience of death.

Physicians will deal with people from a variety of religious or faith traditions. America is increasingly pluralistic. Buddhists, Hindus, Ba'hai, Orthodox, Muslims, and secular humanists are now commonplace in metropolitan or university settings. Each brings a particular pattern of beliefs and favorite texts from scriptures sacred in their tradition. The encounter between people of different religious beliefs need not be a confrontation and should not be in the clinical context. One dimension of respect for the patient is an appreciation for religious beliefs and practices.

Physicians face the need to deal with patients in terms of their faith, which requires a level of comfort with people of different religious expressions and sufficient knowledge of the tradition to carry on an informed conversation. Few things comfort a patient like a physician's knowledge of and appreciation for their culture and faith. The great temptation for physicians with strong beliefs will be to convey disapproval to the patient just because of his or her religious beliefs, which happen to be "different." I have known physicians who felt obliged to resuscitate non-believers just so they would have another chance to be converted. The clinical setting should not be a battle over religion, nor should a physician contribute to a patient's anxiety by challenging or disputing religious opinions. The dying time is not an occasion to mock another person's faith or their comfort in not being a person of faith. Whether the patient is a Muslim, Buddhist, or secular humanist should make no difference to the physician. Each patient will have a particular way of thinking about death, which they find intellectually satisfying and spiritually comforting.

The first task of the physician is to understand and be affirming of the patient. One does not have to agree with them to be supportive of and interested in them. Another critical task is to avoid a form of paternalism or religious imperialism—imposing religious beliefs or insisting on "sharing" your faith to a patient who has not inquired. The task is to develop the ability to deal openly and comfortably with the spiritual needs of the dying in terms of his or her own religious tradition. In so doing, the physician can assume a priestly role and assist the patient as preparation is made for death.

Preparation for Death

End-of-life care in the clinical setting often includes various discussions about and preparations for death. Some people engage in exercises designed to facilitate death. The medievalists cultivated what they called the *ars moriendi*, or the art of dying, exercises for the very old and infirm to prepare for death, commonly practiced in the convents and monasteries. The person mentally and spiritually projected his or her spirit onto the eternal spirit and consciously willed or intended to die. Breathing techniques and prayers were used to enhance the art of giving the self over to God in death. They learned the fine art of "letting go" of this life, and the acceptance of the last and most profound change one would make in human bodily existence.

Judaism, Christianity, Buddhism, Islam, and Hinduism all teach their followers to let go of this life. They are strangers on and to this earth with a citizenship in a transcendent realm. Jews, for instance, may contemplate the meaning of Abraham's leaving Ur for a land about which he knew nothing. He had to let go of where he had been, to forsake cultural comforts, the security of the familiar, and to be ready to face the insecurity of the unknown.

Hindus have rituals of withdrawal *(Vänaprastha)* to be cultivated by the aging and the dying. The theme of self-emptying, or *kenosis*, is found in many religions and was at the center of Jesus' life (Phil 2:7). Older believers have the possibility of detaching themselves from worldly pursuits for the sake of gaining the spiritual gift of peace in dying.

Diognetus once said that Christians "take part in everything as citizens and put up with everything as foreigners. Every foreign land is their home, and every home a foreign land."[14] Aging is and should be a time for letting go—of childhood and adolescence, of young adulthood and middle age, of good looks and youthful energy, of fine health and accumulating wealth, of loved ones and friends, of beloved places and familiar faces, of prestige from social place and business success.

The Physician as Minister to the Soul

Nathaniel Hawthorne's novel *The Scarlet Letter* tells the poignant story of the illicit romance between Reverend Dimmesdale, a young minister, and Hester, a woman who was married to the physician Dr. Chillingworth. The forces of nature won out in the tension between Puritan morals and romantic feelings. The young Dimmesdale violated his oath as a minister and his commitment to moral uprightness. The power of attraction to one another turned out to be far greater than social convention and moral or religious expectations and commitments. He felt terrible remorse as a result

[14] "The Letter to Diognetus," in C. C. Richardson, ed., *Early Christian Fathers* (Philadelphia: Westminster Press, 1953).

of the infidelity involved in their relationship. Chillingworth suspected their relation and took every opportunity to gouge the conscience of both Hester and Dimmesdale. Hester was forced to wear the scarlet A in public to display the shame she bore in private. She bore the letter with dignity in a way that judged the haughty and hypocritical. The guilt and secrecy took its toll on the minister's health, however. He was driven to Chillingworth for assistance, Dimmesdale asked in pain and distress, "You have not, I take it, a remedy for the soul?"

The irony is that physicians do, in fact, have medicine for the soul, though many are ill prepared to assume the role of spiritual counselor as well as medical specialist. Physicians have a special relation with patients that is highly private and personal. The most intimate matters may be shared by the patient with the physician. Among those are spiritual burdens that may take the form of illness not related to physiological but psychological processes or spiritual conflicts. One man had a paralyzed hand, though no neurological problem could be discovered. His history revealed the problem. He was the CPA for a major company, and his boss had insisted he alter the books for a more favorable tax situation. The man was a person of integrity and felt profoundly compromised by the order. His conflict morally was compounded by his financial situation: he was told either to alter the books or be fired.

His emotions so rebelled against the internal conflict that his hand became paralyzed. Once the ethical issue was isolated and resolved, however, the paralysis went away.

Patients facing bad news of fatal illness or the imminence of death may also look to their physician for spiritual solace or guidance.[15] As they seek religious resources of strength, courage, and wisdom,[16] they may want their physician to share in reflections or prayer. Physicians need special wisdom for such occasions, especially when the patient may be of a different religious persuasion. What is required is that the physician be sufficiently comfortable with his or her own religious commitments that other perspectives can be respected and honored. Between Christians and Jews and Muslims, for instance, an enormous amount of biblical material is shared in common. Prayers can be fashioned that appeal to the patient and are still meaningful to the physician. Prayers can be more or less generic,

---

[15] See G. K. Vandekieft, "Breaking Bad News," *American Family Physician* (December 15, 2001).

[16] Paul D. Simmons, *The Southern Baptists: Religious Beliefs and Healthcare Decisions* (Chicago: Park Ridge Center, 2002).

based on the belief that God is a universal reality, no matter the particular religious beliefs of the various groups. Both good faith and professional standards require the physician not use the occasion as an opportunity to try to convert or insult the other. Confrontational religion might contribute to a spike in blood pressure or other adverse health effects. Any interest in converting to another faith tradition should be initiated by the patient. The physician's role is to console and encourage the patient with prayers that tap his or her own spiritual resources.

The patient may also want special spiritual guidance when difficult choices are to be made. Typically, a patient getting bad news will turn to favorite passages of Scripture, engage in prayer, ask for religious counsel, and consult with the family, but the physician has a special and intimate role. The news and the manner in which it is shared may be devastating or it may be conveyed in a caring and compassionate way that the patient can manage with dignity. The physician's discussions with the patient need to be candid but compassionate.

There are times when medical intervention results in burdens that outweigh benefits and decisions must be made as to preferences and appropriate care. The patient may be better off without an aggressive rescue effort, as when the patient will die more comfortably without interventions or invasive procedures.[17] In such cases, spiritual values will largely determine medical choices. A discerning physician will assist the patient in discovering those resources.

A surgeon explored the risks and possible benefits regarding various alternatives facing a seventy-three-year-old man who was diagnosed with a thoracic aneurism. He had full-coverage insurance, but the risk of surgery was considerable. If not successful, the patient would have devastating disabilities. Expenses would eat away the funds of his estate. If he decided against the surgery, on the other hand, he could expect to die comfortably, quickly, peacefully. After further reflection and discussion, the man said to the surgeon, "You could have made a lot of money doing this procedure, so I really appreciate the time you have taken to honestly outline my options. I think the right thing for me is to not have the surgery."

In another case, a woman who was facing bypass surgery had expressed fears of going through surgery. She complained of intense abdominal pain. Examination showed a catheter clotted at the end of the aorta, leading to ischemia of her lower extremities. A nurse suggested she just be allowed to

[17] Joanne Lynn and James Childress, "Must Patients Always Be Given Food and Water?" *HCR* 13/5 (October 1983).

die. The woman was at first refusing surgery, and her family was divided over how to proceed. The doctor calmly helped the patient understand her situation medically. "Yes," he said, "you might die in surgery. But without the surgery you will die an ugly and painful death. You will be facing renal failure; your legs will rot off, and you will spend two weeks dying. If you are going to die, do so in surgery." She decided for the surgery.

Other Preparations

Practical preparations for death can be made. Patients should have an advance directive, for instance, in which they indicate preferences among various options for treatment. If they have multiple organ failure or a massive stroke leaving them unlikely to regain consciousness, perhaps in persistent vegetative state, what would they like done? They have options, of course. They might prefer to have treatment withdrawn and be allowed to die. Or they might prefer full code, which means they want to be resuscitated should they die or they would be maintained on a ventilator to keep them breathing. Renal dialysis would be implemented at some point, as would other procedures. Or they might prefer a limited trial of intervention (LTI), which could specify that aggressive treatment be employed for a specified period of time (one month, six months) in order to evaluate more precisely the effects of the treatments. If there is no return of consciousness and the patient's condition shows no promise of restored health, the treatments can be discontinued.

The advance directive is available online for patients in every state. One can go directly to the office of the attorney general in a particular state or simply search under advance directives and do a general search for your state. You can download it, sign it, have it witnessed, and date it. Such a written statement could have saved a great deal of controversy, delay, legal wrangling, and political posturing in the case of Terri Schiavo. The important thing is that the patient's preferences regarding treatment be known and, insofar as possible, within the limits of the law, those wishes be carried out.

Another item for attention is the need to name a durable power of attorney for health decisions in the event the patient is unconscious or brain damaged and cannot speak for oneself. The proxy or surrogate can be anyone the patient trusts to carry out the patient's own preferences regarding treatments. Discussions with the intended proxy are therefore needed to clarify just what the patient would want. Patients might also consider writing what is called an "Ethical Will: A Legacy of Spirituality."[18] These may be general statements of one's life philosophy and lessons that might have been learned from experiences or studies. People involved in social upheavals (e.g., war, hurricanes, the Civil Rights movement) or who have lived through difficult times such as the Holocaust or the Great Depression could leave a lasting historical record in this manner, or the ethical will may be dedicated to particular people (son, daughter, niece, nephew, or other person or group) in which the person wishes to leave wisdom or counsel for the benefit of the recipient. Obviously, one must be fully conscious for this to be done, but it would be a lasting legacy and a thoughtful gift to those who receive or share its wisdom.

Conclusion

The aging typically come to experience what all people know by mature reflection. We live with the paradox that life is a gift to be treasured, but it is not a second god. There are times when death is a terrible personal threat, but there are also times when we know that death is not the worst thing that can happen. Some ways of staying alive contradict God's will. Both as we get older and during the dying process, our perception of death may change from seeing it as a despised enemy to that of a welcome friend. Death may be seen as a relief to be sought, an experience to be valued, not resisted with every technological weapon available.

Thomas Jefferson once reflected on the meaning of death as he approached the final stages of his life. He noted that there were times he thought of death as an enemy, or as an experience to fear and despise, but there were other times he thought of the relief that would be brought by "the friendly hand of death."[19] He saw the aging process as a matter of "stealing from us, one by one, the faculties of enjoyment, searing our sensibilities,...until satiated and fatigued with this leaden iteration, we ask our own *congé*."[20] He recalled a statement from a very old friend, who was neither a poet nor a philosopher, who said that he would welcome the coming of death.

For Jefferson, living rightly meant preparing for and accepting death. As he put it, "There is a ripeness of time for death...when it is reasonable we

---

[18] Jacob Riener and N. Stampfer, *Ethical Wills: A Modern Jewish Treasury* (New York: Schocken Books, 1983).

[19] T. Jefferson, *Jefferson Himself*, ed. Bernard Mayo (Charlottesville: University of Virginia Press, 1942) 336.

[20] Ibid.

should drop off and make room for another growth. When we have lived our generation out, we should not wish to encroach on another." He had apparently reached a point of equanimity in thinking about the approach of his own death: "I enjoy good health; I am happy in what is around me, yet I assure you I am ripe for leaving all, this year, this day, this hour."[21] The coming of death, he thought, was final proof of the benevolence of the Being who presides over the world.

[21] Ibid., 330.

Chapter 5

Religion, Politics, and a Right to Die

The tragic case of Terri Schiavo in Florida is a disturbing reminder of what is at stake in the debate about the right to die. Between 1997 and 2005, Terri was at the center of a battle between her husband, Michael, who said she would not have wanted to be maintained on life support, and her parents, Bob and Mary Schindler, who maintained she still responded to them. Michael and her parents embody the debate that has raged for years in the United States over the right to terminate nutrition/hydration in a badly brain-damaged patient. Terri was forty-one (2005) and had been maintained in persistent vegetative state since 1990 when she collapsed of a heart attack from a chemical imbalance apparently related to anorexia. She never recovered consciousness.

The right to refuse treatment in such cases was dealt with by the US Supreme Court in a case involving Nancy Cruzan,[1] which had eerie parallels to the contest over Terri Schiavo. The court declared in Cruzan's case that a patient could be removed from nutrition/hydration by a guardian if it were shown that the patient would not wish to be maintained by aggressive medical technology beyond the point at which one could reasonably be expected to recover consciousness.

The tests for withdrawal of treatment were thus focused on two issues: (1) the medical evidence of extensive, irreversible brain damage; and (2) the personal beliefs or best interests of the patient whose values indicated a desire not to be sustained in this fashion.

Between those two tests is enormous room for debate. Doctors sometimes disagree about the degree of damage, families often interpret reflexive actions as cognitive and volitional, political action groups line up to defend their agendas of dignity in dying or the duty to keep alive, and politicians seldom miss a chance to posture for the benefit of their voting constituency.

---

[1] *Cruzan v. Director, Missouri, Department of Health*, US, 88–1503, June 1990.

Governor Jeb Bush's intervention with the Florida legislature to have the gastronomic tubes replaced was regarded by the courts as terribly misplaced and badly abused executive and legislative powers. The governor's actions complicated an already tragic case and simply prolonged the battle with and in the courts.[2]

The Schiavo case in many ways recapitulated an earlier battle in Virginia over the withdrawal of treatment from Hugh Finn, a popular television personality from Louisville, Kentucky. Finn's case involved a battle between his wife, who wanted to allow a natural death, and his parents, who believed that withdrawing treatment was tantamount to murder.

In both the Schiavo and Finn cases, the spouse apparently represented the patient's wishes not to be maintained in a state reminiscent of Karen Anne Quinlan from the 1960s or Nancy Cruzan in the 1980s. Both cases drew national attention and established precedents for similar cases, but withdrawing nutrition/hydration ran up against a vocal and politicized opposition claiming a religious motivation. The debate is not new. It keeps coming around with the same core arguments and points to be considered medically, legally, and ethically. Only the players and places seem to change, not the fundamental arguments.

Religious Perspectives

People of faith are deeply involved in this debate and its outcomes, but they are hardly in agreement about the critical issues at stake. Christians often lead the forces opposing one another, each side claiming the better argument. On both sides, biblical, theological, ethical, and legal perspectives are interwoven with the language of human rights.

This chapter will outline the arguments and isolate what seem to be the primary points at issue in what has become an acrimonious and intractable debate.

Opposition to the Right to Die

Arguments against the right to die are found in writers such as Edmund Pellegrino, Daniel Callahan, David Orentlicher, Richard McCormick, and others. They oppose the notion of a right to die, believing it inconsistent

---

[2] The Florida Supreme Court ruled "Terri's Law" was unconstitutional. The court saw the law as an effort by the governor and legislature to bypass the courts. *The Courier-Journal*, 24 September 2004, A1.

with essential moral and religious values and an attack on the integrity of medicine itself. They represent perspectives informed by the Roman Catholic natural law tradition, but they are not alone.

Evangelical Christians join the informal coalition of those who oppose anything like a moral support for a right to die. Leon Kass, Alan Verhey, Richard Land, and others join the chorus of those who supported Jeb Bush and other politicians who pursue legal measures against actions like those of Michael Schiavo, the parents of Nancy Cruzan, the parents of Karen Quinlan, and the wife of Hugh Finn.

The arguments of those who oppose a right to die can be summarized around certain critical points. The first argument is that claims about a right to die are confusing and meaningless because they are applied to such a variety of procedures or actions. Leon Kass says that what is included is the right to refuse treatment so that death may occur, the right to be killed or to become dead, the right to control one's own dying, the right to die with dignity, and the right to assistance in death. His conclusion is that advocates of a right to die mean that we have a right to become dead even if that requires the assistance of others.[3]

The second contention is that life in any form is regarded as infinitely better than death. Life is thought to be a sacred and absolute value—an unrepeatable gift that is to be respected and preserved under all circumstances. The "sanctity of life" argument emphasizes the "fact" of life or the presence of animation, as the "life" worthy to be sustained and the moral point at issue. The value of life is coupled with a notion of death as "the last enemy" or an evil to be avoided. Thus, life is to be preserved—or death to be resisted—as long as medically or technologically possible.

The third argument is that aggressive measures extensively applied are regarded as the last acts of care and love for the beloved. Family members frequently are comforted by the feeling that "we did all that was possible." Working through grief and loss of love for the person frequently seems to require even extensive and futile treatment as signs of love.

A fourth argument is that hope is a vital factor in the process of patient recovery and an expression of Christian faith in the face of death. Keeping alive the vital life signs makes possible a miracle from God or a medical

[3] Leon R. Kass, "Is There a Right to Die?" *HCR* 23 (January/February 1993): 34–43.

breakthrough making recovery possible. This argument is enforced by dramatic cases of recovery when it seemed the patient was already dead.[4]

Fifth, it is argued that the "pedagogy of death" must not be short-circuited by premature dying. The belief is that there are lessons to be learned during the dying process that can be learned in no other way. Thus, the patient must not be robbed of the final stages of dying (living) lest the benefits that suffering can bring be lost.

The sixth argument is that a right to die would erode the covenant of trust patients enjoy with their physician. Patients must be assured that they will be aggressively treated in their vulnerability and helplessness. Otherwise, the entire fabric of confidence in those to whom we are entrusted would be undermined.

A final argument draws on the notion of the right to life. Philosophers like Kant are cited to argue that a choice to die is a fundamental contradiction—what Kass calls a "nonsensical and nihilistic" claim. In short, the only one who could exercise such a right would no longer be around to enjoy its benefits; the right to die would contradict the right to life because the person would be killing the self that exercises the right.

Excursus: The Question of Authority

Some attention should be given to the wider context in which the debate about a right to die is taking place. At least three major considerations are interwoven in the debate about medical responsibilities and the right to die.

The first is our ambivalence toward medical technology. In 1942, my maternal grandfather died and stayed dead. He was eight-two and thus had lived to realize his "four score years" and two. There was no defibrillator to shock his heart back into action, nor were there any ventilators to provide mechanical breathing for patients in respiratory distress. That machine was not invented until much later. The developments in technology have challenged traditional notions of death and life and when it is ethical or right to declare a patient dead. We need only to be reminded of the shift in the medical and legal definition of death. A generation ago, one was considered dead with the cessation of cardiac and respiratory activity. Now one is thought clinically dead when the brain is dead. The shift was because of new medical technology. Resuscitating patients whose heart has failed is now

---

[4] See H. A. Cole with M. M. Jablow, *One in a Million* (Boston: Little, Brown & Co., 1990).

routinely done in the hospital setting, and defibrillators are available on airplanes and in other public places.

New medical technologies have brought about an era in which matters of life and death are no longer as clear and non-controversial as when no such interventions were available. We are now in that ethical penumbra dealing with matters of life and death in a technological mode. Before the inventions of ventilatory support or gastronomic tubes, we would not be arguing whether Terri Schiavo should be allowed to die or whether her "right to life" required ongoing efforts to support her vital signs. She was certifiably dead in 1990 when she had a heart attack. Her vital signs were restored by artificial means; they were then sustained because of such mechanical interventions.

The ironic truth is, as her case shows, that science is better at prolonging the vital signs than at restoring health and personal responsiveness. Patient autonomy, functionality or consciousness, and intelligent thought are the hallmarks of personhood. Without these a patient is hardly the image of health and vitality that beneficent medicine has in mind. Thus the ambivalence about technology: we want it, but do we not *always* want it. Mechanically supported vital signs challenge our thinking about when it is morally justifiable to declare one dead or allow one to die a natural death and thus withdraw treatment.

A second factor is the crisis in patient-physician relationships and the wider conflict in medical decision-making processes. The issue is one of authority. Who is to decide when death is near but not quite overwhelming the biological or neurological system? Who is the proxy capable of deciding for another, especially when people of intelligence and good faith disagree? Should doctors be in charge of the death decision, or should a lawyer be appointed for each patient during their dying time? Should the spouse or parents of the incompetent adult be the proxy/guardian of choice? The emotional arguments compound the complexity of the decision regarding authority. Should even able adults be allowed to make judgment calls as to when it is ethically justifiable to choose death instead of living under unacceptable conditions?

Third, the conflict and disagreements are at the level of ground-of-meaning beliefs. Profoundly held religious or philosophical beliefs are at stake on both sides of the debate. It is not that one side is religious and faithful and the other is atheist and impious, in spite of arguments sometimes heard to the contrary. Fear and denial often parade as faith when anxiety about death comes into focus.

Perspectives about death and life and whether we have any prerogatives at all in decisions about death are determined by commitments and values by which we define the self and what faith in God and being human are all about.

Religious liberties are thus at issue in important ways. What are the protections and entitlements regarding one's acting on the basis of one's own meaning of life and self-defining beliefs? Such questions permeate our approach to the right to die.

Support for the Right to Die

Significantly different moral perspectives are held by Christians and others who believe in a person's right to die. They can be summarized around five central arguments.

First, human dignity is at issue in decision-making regarding death and dying. There is little, if any, dignity in debility or total incapacitation. Our humanity consists primarily, though not exclusively, in our capacity for reflective choice and personal interactions. In our choices, we define the self and live out the life of faith with all its complexities and ambiguities. Passivity in the face of dysthanasia (ugly death or dying) is a denial of the very capacities that define our creation in the image of God, according to this approach.

Second, love for the patient requires more than simply keeping someone alive. Care extends to comfort and the presence of people to touch and caress us in our times of vulnerability and helplessness. Love accepts my mortality and allows me to die. It does not insist that I remain "alive" simply because it is a technological possibility. Death may be preferable to existence under conditions that deny what it means to be before God. Paul Ramsey spoke of "the countervailing requirements of love," when actions that may not appear to be love actually are.[5] That insight is as true about killing in war as it is in allowing a loved one to die.

Third, the moral weight of patient request or preferences must be regarded with the highest respect. This patient may prefer not to be kept alive. That preference expresses a right, an obligation, on the part of caregivers not to impose treatments inconsistent with the request. Paternalism is tyranny when it ignores the patient as person and insists on treatment based on moral and religious assumptions not held by the patient.

---

[5] Paul Ramsey, *The Just War: Force and Political Responsibility* (New York: Charles Scribner's Sons, 1968) 206.

When Diane asked Dr. Quill for a prescription for a lethal dose she could take in order to avoid the final stages of dying with leukemia,[6] she appealed to his sense of moral obligation to take her seriously as a person. She wanted him to take seriously the medical oath to do no harm or to prevent the harm of being denied a better exit than the one that could be predicted based on her medical condition.

A final argument is that death is not the final victor. Life succeeds even when death intrudes. A Christian theology of death asserts that death, as enemy, has been overcome in Christ's resurrection. Desperate efforts to deny death are idolatrous, since they elevate death to the level of an absolute value. Life is not a second God. Death is to be accepted and at times even invited when life and its purposes under God can no longer be served.

Further, death might be seen as a gift from God. Death is a good gift, in spite of the existential pain and the grief that attends our losing those we love. Death is a sign of God's providential care in both creation and redemption. God has acted to provide relief from suffering and a transition to eternity. What God has provided cannot be an ultimate evil, though it may have elements of evil in particular cases. Christians can find persuasive examples of moral heroes in Scripture who chose death. Jesus, for Christians, is the central theological model and ethical norm. In dying, he said, "Father, into your hands I commend my spirit," thus accepting that the moment had come, and willingly accepted his death.

Further, the Apostle Paul suggested that a choice might be made for death when life had completed its course or it was no longer possible to pursue the central moral and religious projects that give meaning to life. In Philippians 1:20–24, Paul outlines a consistent and coherent argument regarding the grounding in faith that made it possible for him to contemplate the possibility of choosing to die. He began with a strong argument that is basic to a Christian affirmation of the purpose of life and its relation to options about death. His intention and commitment were that he might live courageously in order "to honor Christ through his bodily existence, whether by living or by dying" (v. 20). For him, life's purpose was to serve Christ, and when that was no longer possible, death was to be preferred. Life in the flesh, he said, should be a matter of "fruitful labor."

Paul then added a rather astounding assertion: "which I shall choose I cannot yet tell" (v. 22). He admitted that he was struggling (hard-pressed) with the question. Then he declared that "my desire is to depart (i.e., die),

[6] Timothy E. Quill, "Death and Dignity: A Case of Individualized Decision-making," *NEJM* 324/10 (March 7, 1991): 691–94.

and be with Christ, which is far better (than life while unable to pursue God's will)." His resolution of the question for the time being, at least, was that "to remain in the flesh was more necessary for their sake" (v. 24). Christians, according to Paul, can affirm that "whether we live or whether we die, we are the Lord's" (Rom 14:8).

Affirming such theological points does not settle the issues being confronted in the debate about a right to die, but they do provide an important reminder that opposition to elective death cannot claim to be the *only* Christian position. Paul's struggle with the issue seems to end by affirming what many would regard as a solid basis for a biblical basis for a right to die.

The Continuing Debate: Points of Contention

The debate about one's right to die will continue since the complexity of the issues will hardly go away. There is enough in this debate to occupy the best legal and ethical minds for years to come. Certain questions are central, however, and merit isolating for attention as a way of summarizing and drawing conclusions.

The Nature of Human Rights. First, there are major disagreements about the way in which the notion of human rights is to be construed. Is the right to life a matter of preserving and protecting the vital signs, no matter what, or is it a right to pursue a life (and choose death) consistent with one's own priorities, values, and commitments?

There are many approaches to human rights, of course.[7] Writers do not all mean the same thing when speaking of human rights, or of natural rights. Nor do they agree on what rights are central. We can agree generally that there is a difference in negative rights and positive rights, legal and moral rights, but which is which? Of critical importance is whether a right to life includes a right to die, or whether they are mutually exclusive.

Further, there is a question as to whether rights language is consistent with religious approaches to moral issues as developed in the Christian heritage of thought, especially that which centers on the biblical revelation. Human rights language dominates most discussions in medical ethics, and Christians should be comfortable with the compatibility of rights with biblical insights and moral guidance. The biblical revelation is consistent with and provides warrants, backings, and supports for human rights. The

---

[7] See Barbara MacKinnon, *Ethics: Theory and Contemporary Issues*, 2d ed. (Belmont, CA: Wadsworth, 1998) 75–80.

1948 UN Declaration of Human Rights, for instance, is a document worthy of religious support and commitments. Drawing parallels or blending the moral concerns of human rights with those of Scripture seems entirely acceptable.

Human Rights and Universal Norms. The great value of the human rights argument is that it appeals to universal norms, whether or not everyone accepts them. No one is excluded from the moral protections encompassed by human rights, but writers like Callahan and Orentlicher[8] appeal to universality in ways that are both inclusive and exclusive. Beneath their approach is a claim to truth *no matter what others might think*. If others disagree, then the truth is to be imposed by law or structures of power since the claimant has laid hold of the natural law. The human right to life, for instance, is construed as binding even on the beneficiary of that right. The central model is law, not ethics. The natural law, or law of the universe, or of God, is the source of the right being protected and obeyed, according to this view.

Are Rights Inalienable?

Those opposing a right to die argue strongly that the right to life is inalienable. The right is bestowed as a binding obligation. It cannot be bartered or voluntarily surrendered by the holder, regardless of circumstances. The argument adds that the civil law should embody this moral constraint.

Collapsing moral claims into legal norms poses a major problem, of course, but it is done without blinking in the Catholic tradition of moral law.[9] Orentlicher argues that a person has no permission to waive or voluntarily consent to surrendering a right once bestowed. He says that one cannot give away the right to vote or voluntarily enter into slavery or commit homicide.

Even so, proponents disagree as to whether (1) one might forfeit the right to life by taking another person's life (thus approving war and capital punishment), and (2) whether refusal of treatment would violate the right to

---

[8] Daniel Callahan, "The Immorality of Assisted Suicide," in *Physician-assisted Suicide*, ed. R. Weir (Bloomington: Indiana University Press, 1997), and David Orentlicher, "Physician-assisted Dying: The Conflict with Fundamental Principles of American Law," from *Medicine Unbound: The Human Body and the Limitations of Medical Intervention*, ed. Robert H. Blank and A. L Bonnicksen (New York: Columbia University Press, 1994).

[9] See the Vatican document *Respect for Life*.

life. Orentlicher supports both capital punishment and war, while Callahan denies that either is a Christian possibility.

One might also wonder whether the moral issue of killing does not involve necessary distinctions between motives such as malice and mercy. Christians are typically more comfortable arguing for killing as a way to retaliate in kind than they are discovering that mercy might also require assistance in dying (another type of killing). Orentlicher concedes that the right to life means at a minimum not to have one's life taken by another. Surely he should have qualified that statement by saying "taken by another with malice and aforethought," or without the foreknowledge, consent, or request of the dying.

Assistance in dying, on the other hand, may be required by mercy and what Ramsey called "the countervailing requirements of agape." The classic case of the man dying in a fiery crash who begs a highway patrolman to shoot him illustrates the moral rigidity, if not blindness, at the heart of the "inalienability" argument.

A right is a moral construct or "myth" designed to express the liberties, entitlements, privileges, or protections belonging to individuals, and the obligations imposed upon others to provide such benefits and/or the restraints necessary to protect the exercise and expression of protected freedoms by the individual (or group).

The origin of rights was in the struggle between persons in positions of power that took advantage of the less powerful and exploited the powerless. Issues of power and authority are at the heart of the moral concern. "Rights" are invoked by individuals and groups who perceive that their entitlements are threatened by an unjust exercise of power and/or authority over them. Rights claims a superior moral ground against the pragmatic or ideological claims of authority or power by various institutions, groups, or individuals.

The first concern for human rights was political in nature. The king was the problem. He claimed a "divine right" to govern as he saw fit, even to the point of dictating conscientious religious convictions and that all citizens and their property belonged to him. He could thus require submission and subservience of all subjects no matter the degree of humiliation or injury he might impose economically or personally. After all, the king claimed to own the life of all citizens. Whatever the king did was a corollary of his special status as the appointed of God.

The second great claim on behalf of human rights was against the imperial church. The Pope claimed divine prerogatives over the destiny of the soul and that only he could manage the affairs of faith before God. It took the insights of the Reformation and Baptists like Richard Overton to remind the king of the prerogatives of the individual conscience in matters of faith, an insight with enormous relevance to the current debate.

Claiming a human "right" is a way to leverage or influence the prerogatives of the person against any effort to infringe upon or limit liberties important to the individual. The question is the value system of the holder of rights, not the imposition of rights claimed by an alien and oppressive power whether king, pope, physician, politician, minister, or neighbor. The problem with the right to die is that so many others are interested in their interpretation of rights that entitles them to take away liberties the dying hold crucial and central to personal well-being.

My commitment is to the belief that each person has equal access to God, thus making each person an authority unto his or her own self in matters of conscience or belief. Personal autonomy, not heteronomy, wins out in human rights. The individual is the focus of value. He or she is not to be violated in body, mind, or spirit.

What a Right to Die Permits or Requires

The emphasis and focus on the individual assure that the debate about human rights will continue without interruption. Important segments of the medical and social community are more inclined toward a hierarchical structure of decision-making that emphasizes the importance of authority related to position, not the individual, as the locus of value and authority in decision-making. This issue has permeated the world of medical practice as much as the world of religion and politics. We need only be reminded that earlier texts in medical ethics emphasized the physician's prerogatives and diminished or discounted the importance of patient preferences, based on values and commitments, which are both moral and religious.

The tension between these competing perspectives is also found in the Supreme Court decision regarding assisted dying in the case involving Dr. Quill. The court stated emphatically that one has no right to an assisted death. They were apparently attempting to protect the physician's prerogative not to engage in practices for which they have no sympathy. At most, the court said there was no positive right to assistance, but it did not address the murkier arena of whether there is a negative right regarding death decisions. It is also possible that they ignored that arena because it is a moot point where the law is concerned. People will still make their own decisions, no matter what the court says.

Another point to be made is that there is a developing body of law that seems implicitly to affirm a right to die. Those decisions have been developing now over a thirty-year period. The fact that the Supreme Court allowed the "death with dignity" law to stand in Oregon is both a recognition of the prerogatives of the various states and an implicit recognition of a human right to die.

That thirty years of accumulated wisdom and case law make Kass's statement that the right to die means many things[10] perfectly true and understandable. The fact that he does not like the idea of a right to die underscores my point that individual perspectives are still in tension with authoritarian approaches. What Kass finds objectionable or unacceptable seems to belong to the complexity and ambiguity that accompany death and dying. The right to die does in fact apply to cases involving the withdrawal of treatment, as in the Schiavo case; to cases of assisted suicide, as Oregon now allows; and to allowing a person to die a natural death instead of being resuscitated as many times as technologically possible.

A right to die also applies to suicide, especially in those cases in which one is dying from a terminal illness, and it applies to those persons who are being sustained after resuscitation but should have been allowed to remain dead. Only a misguided and imperialistic ethic would argue that every person should be resuscitated no matter the circumstances of death or the beliefs and values of the dying.

Once the issue of the right to die is raised, it is difficult to draw the line with a particular set of actions and say it pertains to this but not to all others. The distinguishing feature is that a right to die relates to actions that intentionally and knowingly intervene in the dying process in order to bring about a more merciful and peaceful resolution to an otherwise agonizing and personally distressing process.

By my reading, certain elements of a right to die are now implicitly and tacitly recognized in law, medicine, and ethics.

A Right to Die a Natural Death. Do not resuscitate orders, for instance, seem to acknowledge a right to die a natural death. Such orders are widely used and are standard medical practice. They are accepted as both a negative and a positive right. When a patient requests that no attempts be made to resuscitate in the event of cardiac or respiratory arrest, and physicians agree that the prospects for healing are minimal at best, the order may be entered on the chart. At that point, any effort to resuscitate a patient constitutes battery or negligence.

---

[10] Kass, "Is There a Right to Die?" (see his note 3).

But if one has no right to die, all technological efforts to prevent death would be morally mandated. One physician was resuscitated ninety-six times, during which he continually begged his colleagues to let him go. He had a right to be left alone, to be allowed to die, to not be mauled and manipulated by well-intentioned but misguided physicians. He had a right to die.

A *Right to Refuse Treatment.* Medical interventions one finds unacceptable may also be refused or rejected as long as one is fully informed of the predictable outcomes. The choice is between continuing to "live" as the object of aggressive medical interventions and dying on one's own terms. Deciding against open-heart surgery may well be a decision to die quickly and painlessly, living life to the full meanwhile, instead of chancing death on the operating table or after a lengthy time on the respirator.

We respect the decision by adult Jehovah's Witnesses to refuse blood products based on their belief that doing so would jeopardize their life in eternity. We may disagree with the theology and the biblical interpretations upon which such a decision is made, but we respect their right to act upon religious opinions and beliefs.

When James Michener refused any further dialysis, he was acting out of profound beliefs about life and when it is right to accept death. He "stuck out his tongue at the encroachments of advancing technology and freely and responsibly chose to die,"[11] as Ramsey so memorably put it.

A Choice among Alternatives. The right to die also means one may choose among alternatives regarding the manner in which one is to die. There are many ways to die, of course. One alternative we do not have is not to die. We will die; the only question is how or under what circumstances.

Most people die without having made any choice among possible alternatives. One may decide to live until one dies of whatever circumstances that might finally overwhelm the biological and neurological systems that have sustained one's bodily and personal existence, or one might be taken out by an accident that kills suddenly and without warning. A man may be killed in a logging accident, or a young woman in an automobile crash. Their life was snuffed out instantly. They had no choice in the matter. Thus, one may live until one dies, thinking little of choices among options.

But that approach to death is something less than making a choice about death. It is a choice to avoid choosing. That approach is certainly one's right—no one would suggest that the heroic life must be chosen over

---

[11] Paul Ramsey, *Fabricated Man: The Ethics of Genetic Control* (New Haven: Yale University, 1970) 150 (Ramsey's emphasis).

the acquiescent, but one does have a limited right to choose among alternatives.

There are times, for instance, when a choice among possibilities offers itself as a better way than others that can be anticipated. Saul, the first king of Israel, made such a choice. He faced capture, imprisonment, and torture by the Philistines. He was mortally wounded in battle, but he would die slowly, giving the enemy ample time to do things to him involving torture and humiliation. The enemy would take great pleasure in gouging out his eyes and placing him on display as the spoils of war. His God would be mocked and his people despised, but he had a choice. He could choose to die quickly at the hand of his trusted and beloved armor-bearer, or by his own hand, by falling on his sword. (Compare the stories in 1 Sam 31:4 and 2 Sam 1:10.) He exercised a "right to die" in a way he found preferable to having the Philistines decide the final manner of his death.

When Diane chose to die with a lethal overdose, she was choosing among two or more alternatives, all of which were undesirable. She had been diagnosed with acute myelomonocytic leukemia. Leukemia is a terrible way to die; in its final stages, the person hemorrhages from every orifice of the body. She knew the problems of fighting cancer since she had been through it before. She had taken the full course of chemotherapy and defeated vaginal cancer. She was a fighter.

She had also overcome alcoholism, but now she faced a bitter choice. She could choose the chemo for leukemia, which carried with it a twenty-five percent chance of recovery, or accept death as leukemia dished it out, or choose what she considered a more dignified way. With her physician's assistance, she died of an overdose.[12] She chose among options. Facing death squarely in the face, she chose one manner of death to another she found entirely predictable but less acceptable.

Bad Outcomes and an Intolerable Existence. Finally, one has a right to die should an intolerable existence be imposed by a bad outcome from medical interventions. There are cases in which a patient has been resuscitated and the vital signs restored but without consciousness or any possibility of recovering self-awareness. Those in persistent vegetative state, such as Terri Schiavo, are examples of medical interventions with less than desirable outcomes.

The list of patients subjected to a life of futility in which they know neither themselves nor anything of their situation is getting terribly long. The extensive numbers (as many as 30,000 nationwide) are a testimony to

[12] Quill, "Death and Dignity," 691–94.

the technological takeover of medical care. Medical science is far better at keeping alive than at healing the brain-damaged patient. The courts have heard a number of cases, and the media has held their stories up for examination and reflection. Several have become virtually household names: Karen Anne Quinlan, Nancy Cruzan, Sue DeGrella, and now Terri Schiavo.

Add to this list of tragic names the story of the young woman who had a history of attempted suicides. For profound but undiscoverable reasons, she was bent on self-destruction. Finally, she took a shotgun to herself and blew away the side of her head. She had finally killed herself, but ER technicians did their magic, and she was resuscitated. She was left "locked in"; she was not fortunate enough to be brain dead and thus to know nothing. She knew but could do nothing but groan in an indescribable misery. She is quadriplegic and thus will never be able to attempt suicide again. By some readings, she is the quintessential example of not having a right to die—certainly she can exercise no powers against her biological existence.

Medicine did its best, but she experienced what is arguably the worst possible outcome of aggressive medical intervention. The purpose of medicine is to restore health and vitality, in short, to restore personal autonomy. The person as person with all her or his capacities for life and living and choosing is the object of care in humane and moral medicine, but that purpose is severely tested, if not contradicted, when patients wind up in a limbo between life and death.

Callahan argues that the patient must accept even the very bad outcomes of aggressive medicine. He consigns the patient to an intolerable existence and would tie the hands of compassionate physicians who would assist their demise. Callahan even denies that physicians have any (moral) duty "to relieve that suffering *which was caused by their treatment*." [13] He would leave the patient in a purgatory of medicine's own making. For him, there is "no call for reparation when (medicine) fails, nor is unavoidable pain and suffering to be understood as a failure of medicine."[14]

The reason he gives is that the treatment was an effort to restore health and thus was a morally praiseworthy action. The physician had good intentions, even if they did not work out well. He pulls back from saying what might logically and morally follow by arguing that the physician has an ethical obligation to correct the problem that has been created. Without the aggressive if not desperate interventions in the ER the patient would have

---

[13] Daniel Callahan, "Immorality," 212 (my emphasis).

[14] Ibid.

died (and actually had died) a natural death. The body and/or brain was so assaulted by disease or injury that death would have overtaken the person and mercifully brought relief. The medical intrusion prevented a more desirable outcome and caused a morally intolerable situation for the patient.

James Rachels points out that the operative doctrine behind arguments like that by Callahan is that the patient may be allowed to dehydrate, wither, and waste away, but may not be given a lethal dose. This amounts, he says, to an irrefutable cruelty.[15] Nietzsche once said that "there is a justice according to which we take a man's life, but there is none whatsoever when we deprive him of dying: this is only cruelty."[16]

There is neither comfort nor hope when survivors are told there is nothing else medicine can do for them when the patient is left alive but only on life support. Of course there is more that can be done. Bill O'Reilly, of FOX News, said that the right thing to do for Terri Schiavo was to give her an injection to assure a quick and painless death. He argued that condemning her to die by starvation was both a miscarriage of justice and a mismanagement of medicine.[17]

Just what might be desirable under such circumstances is a matter of trust between physician and patient, the very thing Edmund Pellegrino[18] and the Supreme Court[19] say is crucial and central to medical ethics. Patients should engage in the intellectual and spiritual exercise of filling in the blanks to the statement, "I want to trust my physician to…." And those possibilities should be discussed openly with their physician. Pellegrino, Callahan, Kass, et al., fear we cannot trust our physician if we allow assisted suicide.

Bad outcomes from aggressive medical intervention seem to make the real question whether we can trust the physician not to leave us in the limbo between life and death, where we are neither certifiably dead nor demonstrably alive. There is something else that can be done in such cases that is far more humane and consistent with the goals of medicine and the

---

[15] J. Rachels, "Active and Passive Euthanasia," *NEJM* 292/2 (January 9, 1975): 78–80.

[16] Frederich Nietszche, *Human All Too Human: On the History of Moral Sentiments* (Stuttgart: Kröner Verlag, 1959) aphorism 88.

[17] Bill O'Reilly, *The O'Reilly Factor*, March 31, 2005, http://mediamatters.org/items.

[18] Edmund Pellegrino, "Euthanasia as a Distortion of the Healing Relationship," in *Controversial Issues in Bioethics*, 4th ed., ed. T. Beauchamp and L. Walters (Belmont, CA: Wadsworth Press, 1994) 483–84.

[19] *Vacco, Attorney General of New York, et al. v. Quill et al.* (June 26, 1997). (Majority opinion by Chief Justice William Rehnquist. See Appendix VI.)

purposes of human life. If we have the temerity to intervene in the death process, to restore life at a subsistence level, we should also have the temerity to intervene to assist patients to die when there is no hope of recovery.

That is especially true when one's life is patterned and being shaped for life in eternity. Believing that life is a transition to a fuller life is an important factor in any Christian thought about death. The Apostle Paul expressed it strongly: "If for this life only we have hope, we are of all people most miserable" (1 Cor 15:19) and "Whether we live or whether we die, we are the Lord's" (Rom 14:8).

Medicine is at its best when persons with a life-threatening illness or injury are assisted to recover sight, thought, movement, and purpose, but it is at its worst when it steps aside and refuses to accept responsibility for terrible outcomes. In the case of the young suicidal woman, any further responsibility for her well-being was denied. "We have done all we can for her" is an excuse offered as one runs from the burdensome responsibility of taking further actions to relieve the patient of a miserable existence. What was done was undoubtedly intended as a moral action, but the outcome was anything but morally acceptable. The heroic interventions did nothing for her except return her to an existence that was even worse than her insufferable life before the attempted suicide. Further, the interventions took place after she had died. She had a right to stay dead, though physicians in the ER were unaware of her former wishes. In her suicide, she had delivered an emphatic advance directive that should have been honored, howbeit with deep regret. Now she is consigned to an existence no rational human being would choose for oneself. She was placed in prison with no hope of escape. Beneficence took on an ugly face.

Conclusions

The contentious debate about a right to die will undoubtedly continue. The political atmosphere is so poisoned with invective that civility has been lost in the public square. Even so, the courts have managed a rather consistent support for a limited right to die, in spite of harsh criticisms from the religious coalition on the right, but such issues should not be subject to a majority vote. On that point I agree with people like Callahan and Kass, but on opposite sides of the issue. They see the question in terms of a clear and definitive line between killing and allowing to die. They would erect strong legal and moral barriers to any decision in favor of helping a person die. Life—no matter what its presenting face—is to be preserved inviolate.

The issue seems far more complex, and answers must be far more nuanced than is supposed by Callahan, Kass, and others who oppose the practice, it seems to me. Caution is clearly in order whenever we undertake to assist a person to die, but the traffic light is yellow, not red. Proceed slowly and cautiously while marshaling all the wisdom of medicine, law, ethics, and religion that can possibly be garnered, but let no single group dictate the terms under which everyone must live while dying. The right to die is far too important to leave to the courts or to the undying strong who may dictate the terms under which our final days must be lived. Paternalism has many faces, one of which is condescending toward and contemptuous of those who see reality in a way almost diametrically opposed to theirs.

The right to die is highly personal. I have a right to die and stay dead, that is, not to be resuscitated against my will. I have a right before God to claim that those who placed me in limbo will accept responsibility for relieving me of that imprisonment of mind and body. Neither physician nor clergy can or should dare displace the role of God in the believer's life and understanding of when it is right to die and when to stay alive. Certainly any declaration of God's will in abstract and absolute terms should be resisted in the name of common sense and the demands of mercy, not to mention the contrary perception of what God's will actually is. The believer will resist such paternalism in the name of good faith while welcoming the supportive wisdom even of those who disagree with us. The hope is that we can enter a covenant of mutual responsibility with our physician and friends as we enter the twilight zone of imminent death.

Chapter 6

FAITH AND THE FUTURE OF PHYSICIAN-ASSISTED SUICIDE

Jack Kevorkian, the California pathologist who mounted a public crusade for what he regarded as more rational and humane laws regarding assisted dying, for a time was the central figure in the public fascination with and widespread interest in physician-assisted suicide (PAS). He chose Michigan as the staging ground for his efforts since it had no law against assisted suicide, helping at least 130 people to commit suicide, either by breathing carbon monoxide or by triggering a lethal injection for themselves.

The first was Janet Adkins, a thirty-nine-year-old woman who had been diagnosed with Alzheimer's. Another was a physician, Ali Khalili, sixty-one, an oncologist suffering from bone cancer. Going to Kevorkian, he said, was "a political act." He wanted not only to affirm the work of Kevorkian, but to support the campaign to make physician-assisted suicide legal. Still another was a psychiatrist from Wilson, North Carolina, who was dying with colon cancer.

Public reaction revealed considerable discomfort with Kevorkian and his irascible, crusading style, but the public was unsure of what, if anything, he was doing wrong. In May 1994, Kevorkian was acquitted of three different charges that he violated Michigan's law against assisted suicide. The legislature had hastily passed the law to prevent Kevorkian's work with patients wanting to die. Whether the acquittals were a miscarriage of justice or the genius of the jury system at work is a judgment that deeply divides the medical, religious, and political communities.

The ethical question about assisted suicide is much larger than attitudes toward Dr. Kevorkian, of course. Other physicians have also assisted patients to commit suicide. Kevorkian was the most visible and outspoken physician associated with the practice and the ethical questions that attend that debate, but he is by no means alone in pushing the issue. Are such physicians a friend or foe to moral medicine? Was Kevorkian's suicide machine a symbol of a moral horror that should enrage the public and be outlawed in every state legislature, or was it a welcome sign of hope in the midst of a technological takeover at the end of life? What of the future? Will other forms of assisted death become commonplace as we enter a new era of medicine?

The purpose of this chapter is to examine the public debate about physician-assisted suicide and to explore the moral issues at stake. Of interest are both the challenges being made to professional medical ethics and questions pertaining to religious perspectives on elective death.

Kevorkian: Ethics and Public Action

Opinion polls indicate widespread support for physician-assisted suicide in the United States. A number of people believe Kevorkian made an important contribution to the social debate by pushing an important issue in medicine, ethics, and law. Such support may indicate that a major paradigm shift in medical practice is taking place.

Medicine is facing a crisis of confidence. The public perception is that basic human values have been overshadowed or neglected in the commitment to preserve life under any circumstances and regardless of the consequences to the patient, the family, or society. The signs of significant support for assistance in dying seem everywhere apparent.

In 1994, voters in Oregon overwhelmingly passed Measure 16, a "Death with Dignity" act, which allows physicians to prescribe, but not administer, a lethal dose of medicine to a terminally ill patient. The US Supreme Court refused to grant an injunction against its implementation even after its decision in *Vacco v. Quill* (1997), which held that there is no constitutional right to assisted suicide and that states may prohibit the practice. By its deference to Oregon voters, however, it was also saying that assisted suicide may be approved by the states.

The Society for the Right to Die (formerly, the Hemlock Society) strongly supports such laws. Members of the society are committed to the proposition that suicide is a human right and that medical science has a moral responsibility to assist competent patients in their final passage.

A number of physicians have also indicated their belief that it is time to embrace physician-assisted suicide as part of a more realistic and morally responsible medical ethic. A twelve-member panel of the American Medical

Society declared it ethically acceptable to prescribe pills, knowing the intent of the patient is to use the pills to end his or her own life.[1]

Dr. Timothy Quill told his story about assisting Diane,[2] a thirty-four-year-old woman diagnosed with acute myelomonocytic leukemia. She had earlier fought and recovered from vaginal cancer and alcoholism. When given the prognosis for cure, considering the standard modalities of treatments for her type of leukemia, however, she refused further treatments. The twenty-five percent rate of survival for patients who make it through the first round of chemotherapy was daunting to her in light of her prior experiences with cancer. In addition to the low odds in her favor, she considered the pain, loss of control of bodily functions, and total personal impact chemo would have on her. She was, according to Dr. Quill, extremely articulate, strong-willed, and determined not to lapse into total debility and unmanageable pain.

Dr. Quill, a former director of a hospice program and a member of the Society for the Right to Die, supported her right to make the decision. Because he knew her well, found her extremely well informed about the disease and its prognosis, and strongly in touch with her own feelings and beliefs about life and death, he entered an agreement with her. He would give her a prescription for barbiturates, if she would agree to call him and talk further before she actually took the pills. Meanwhile, he would continue with her to provide comfort care and pain management.

A year later, Diane called. She had decided that the time had come. Her husband and adult son, both of whom had shared in the discussion from the beginning, supported her decision. Dr. Quill went to her house to visit with her for the last time. After talking with her, Dr. Quill, Diane's husband, and her son left her alone in the house for a sufficient time for the ingested pills to have their lethal effect. On returning to the house, they found Diane dead on the couch. She was wearing her favorite housecoat. Dr. Quill called the medical examiner and reported that she had died of acute leukemia, which was listed as the official cause of death.

Reporting the exact details of her dying would have led, Quill thought, to an unnecessary investigation by the coroner, and perhaps subjected him, the husband, and the son to charges of assisting a suicide. That process

---

[1] See their statement in S. Wanzer et al., "The Physician's Responsibility Toward the Hopelessly Ill," *New England Journal of Medicine* (NEJM) (March 30, 1989): 844–49.

[2] Timothy E. Quill, "Death and Dignity: A Case of Individualized Decision-making," *NEJM* 324/10 (March 7, 1991): 691–94.

would have been both unnecessary and morally unjustifiable, in his judgment, adding to the tragedy already involved. In most states, his action would be legally vulnerable (as in Kentucky and New York).

There is not a great deal of difference between what Dr. Quill did with and for Diane and what Dr. Kevorkian did with so many patients on so many occasions, of course. The patient actually administered the lethal dose. The doctor's assistance is a major step removed from the act that brings about death, but there are two rather prominent differences between Kevorkian's actions and Dr. Quill's description of his work with Diane: (1) Kevorkian remained with the patient, and (2) he made a point of the issue in a way to draw publicity. He was a crusader for legal reform and thus went public with a type of civil disobedience reminiscent of Martin Luther King Jr. Dr. Quill, on the other hand, showed a becoming and professional reticence by refusing to give the issue a circus-like atmosphere. He came forward in professional circles and attempted to place an important issue on the agenda for extended discussions about ethics in medicine.

When Kevorkian crossed the line between assisting persons to take their own life and actually injecting a patient with a lethal drug, he ran into serious legal trouble.[3] Kevorkian made a videotape of his action on behalf of Thomas Youk, a patient suffering from amalytropic lateral sclerosis (ALS), popularly known as Lou Gehrig's disease. ALS is a disease of the neurological system, which slowly deprives the patient of any ability to move arms or legs and finally stops their ability to breathe. Kevorkian had clearly procured the consent of Youk, as indicated on the video recording. He then proceeded to inject the drug into a shunt that had already been placed in the patient's arm. Youk died quickly and quietly.

Actually injecting a person in order to bring about death is considered homicide in every state and in most Western countries. Exceptions are found in Belgium and the Netherlands.

Objections to Physician-assisted Suicide

There are staunch defenders of a more traditional medical ethics who defend both an absolute moral distinction between killing and allowing to die, and who fear the terrible consequences they envision should PAS be legalized.

---

[3] Kevorkian was found guilty of murder and sentenced to fifteen to twenty-five years in prison in March 1999. He applied for parole on grounds of poor health in spring 2006. He was released June 1, 2007.

Edmund Pellegrino. For instance, in a provocative and hard-hitting article, Dr. Edmund Pellegrino, of Georgetown University, decries and laments the current assault on what he calls "traditional medical ethics."[4] The challenge, he says, is coming from anti-principlism and moral skepticism.

Pellegrino champions and defends an approach that appeals to Hippocrates, Aristotle, and Aquinas, he says. He advocates what he calls "virtue ethics," with an emphasis upon such central principles as beneficence, non-maleficence, and confidentiality, and prohibitions against abortion, euthanasia, and sexual relations with patients.[5] He argues the Aristotelian virtue of *phronesein*, or practical moral wisdom, which he thinks provides the anchor for medical ethics. As he sees it, *phronesein* enables the physician to discern the right and good thing to do in particular circumstances.

Pellegrino admits that medical ethics is undergoing radical changes. "The metamorphosis that has already been so rapid and profound," he says, "is far from complete."[6] He is unwilling to predict the precise shape of the medical ethics of the new millennium, but he is deeply involved in efforts to preserve traditional approaches and principles.

He fears, for instance, that acceptance of physician-assisted death will result in the erosion of trust a patient should be able to have in the physician. Physicians should be counted on always to sustain life, he argues. If ever they begin to help patients die, the patient will not be able to trust themselves to professional caregivers. For him, the purpose of medicine is to restore health when possible and to enable the patient to cope with the desirability of death when cure is not possible.[7]

Pellegrino is right to emphasize the fact that the practice of medicine is grounded in trust. The patient is vulnerable and likely powerless to be cured without medical interventions. The process of assisting a patient to recover health or defeat disease depends to a considerable extent on the good will and character—the moral virtue—of the physician. The "physician must seek to heal, not to remove the need for healing by killing the patient. When

---

[4] Edmund Pellegrino, "The Metamorphosis of Medical Ethics: A 30-year Perspective," *JAMA* 269/9 (March 3, 1993): 1158–62.

[5] Ibid., 1159.

[6] Ibid., 1161.

[7] Edmund Pellegrino, "Euthanasia as a Distortion of the Healing Relationship," in *Controversial Issues in Bioethics*, 4th ed., ed. T. Beauchamp and L. Walters (Belmont, CA: Wadsworth Press, 1994) 483–84.

euthanasia is a possible option," Pellegrino charges, "this trust relationship is seriously distorted."[8]

Other arguments Pellegrino advances are that any form of elective death devalues the quality of the patient's life, or reflects a desire to conserve social resources. He believes these processes will subtly coerce patients to accept euthanasia (sic!) rather than to become a "burden" to society.

Another objection he raises focuses on clinical imponderables. Diagnoses and prognoses are not always correct, he rightly says. That problem is compounded when patients plead for release or the physician is frustrated or emotionally spent.

Pellegrino also fears that physician-assisted suicide will have a deleterious effect on the physician's psyche. He says that doctors will become desensitized to the loss of life, that they will participate in the objectification of death and dying until no sense of loss will be felt when people die. For him, the nature of the healing relationship requires proscribing any form of active assistance in dying. His final argument is to appeal to a common aphorism: "An immoral act does not become moral because it is common practice."

Arthur Dyck of Harvard Divinity School adds to these arguments the notion that assisted suicide will lead to a loss of community. Medicine, he says, is a moral community, and the practice of euthanasia (sic!) is detrimental to the welfare of patients and the integrity of society. It is, he says, "a socially destructive option."[9]

Allen Verhey, of Hope College, objects to decriminalizing physician-assisted death, believing there is a need to prevent what he regards as predictable and inevitable evils should society allow the practice. Maximizing freedoms certainly "expresses a culture that values autonomy, independence and control," he says, but it has the negative effect of placing new pressures on those who are suffering and on our attitudes toward them. "The effect of (legalizing PAS)," he says, "may be to make it more difficult for the sick and suffering...to refuse the option of death, harder to justify their existence."[10]

The American Medical Association, the largest physician's organization in the United States, also strongly opposes physician-assisted suicide and

---

[8] Ibid., 483.

[9] Arthur Dyck, "Physician-assisted Suicide—Is It Ethical?" *Harvard Divinity Bulletin* 21/4 (1992): 16–17.

[10] Allen Verhey, "Choosing Death: The Ethics of Assisted Suicide," *The Christian Century* (July 17–24, 1996) 719.

entered an *amicus* brief opposed to Dr. Quill and others in the case considered by the Supreme Court. The AMA asked the court to protect physicians from the burden that would follow from assisting patients to end their lives. Other arguments were also given, one of which was the notion that there is an absolute moral distinction between killing and allowing to die. The latter is acceptable morally and medically, said the AMA, but the former is forbidden by the ethos of medical practice that is committed to preserving or extending, but never to taking, life.

Further, said the AMA, intention or motive is important morally. Shortening a patient's dying process by giving an injection to control pain that unintentionally suppresses respiration is one thing, but to inject a patient with pain-killing drugs *intending* the death of the patient is quite another.

Finally, the AMA statement argues that the medical mandate to *do no harm* is at stake. The principle of non-maleficence requires never cooperating with death, or assisting a patient to die more quickly and painlessly, in its judgment.

Vacco v. Quill. The Supreme Court decision in *Vacco v. Quill* basically followed the AMA's arguments. It rejected the findings of both the 2nd and 9th Circuit Courts of Appeal that had upheld a right to die under the equal protection clause of the Fourteenth Amendment.[11] Instead, said the Supreme Court, the equal protection clause creates no substantive laws, only requiring that no person be denied the protection of the laws. Thus, New York (or any other state) may ban assisted suicide. Anyone may refuse unwanted medical treatment, but no one is permitted to assist a suicide. The distinction between withdrawing treatment and assisting suicide was regarded by the court as "both important and logical," since it "comports with the fundamental legal principles of causation and intent."[12]

The Supreme Court actually skirted the question as to whether a patient has a constitutional (negative) right to commit suicide. Its interest was more narrowly focused on whether someone (physician or other) might be legally obliged to assist. The court said strongly that patients do not have a (positive) right to demand assistance from physicians, nor do physicians have an obligation, either moral or legal, to provide such assistance.

---

[11] See David McKenzie, "Church, State and Physician-assisted Suicide," *Journal of Church and State* 46/4 (Autumn 2004): 787–809, who argues that Justice Reinhart of the 9th Circuit rightly perceived the danger of imposing restrictions based on religious assumptions in violation of the First Amendment (p. 797).

[12] *Vacco, Attorney General of New York, et.al. v. Quill, et. al.* US, June 26, 1997.

Other arguments can certainly be discerned in the debate. The concerns and fears they express have an internal logic that deserves respect and serious reflective attention. Those who do not accept the arguments set forth will need, at a minimum, to present a more cogent and plausible point of view, while dealing with the perceived flaws and inadequacies of the arguments rejected.

Without responding directly to these arguments,[13] I propose to broaden the debate by setting forth what seem to be the main messages in the widespread support for physician-assisted suicide. The "message" is far more complex than a rational argument or even the ability to unravel or counter the logic of those who so strongly oppose physician assistance in dying.

The Challenge of Ethics and the New Medicine

The movement to support physician-assisted suicide seems even more powerful than its arguments. Perhaps for such reasons it is an irresistible movement with factors that are both desirable and thus inevitable, and some that are less than desirable but nonetheless inescapable. At least seven morally significant claims are being made in support of physician-assisted suicide. They reflect a combination of symbolic action and ethical insight.

In certain ways Pellegrino, Dyck, Verhey, and others represent and defend an approach to medical ethics that is being challenged as to its ethical adequacy and historical accuracy. Is the image that physicians have never assisted the dying true to experience and history, or is such a picture based on an idealized or mythical past?

Physician testimonies indicate that physician-assisted suicide is hardly a recent development or some passing fad that will quickly pass. Through the centuries, physicians have facilitated death in the name of both good morals and good medicine. Remembering the history of medicine rightly is one path accurately to portray the ethics of the profession.

For instance, Francis Moore recalls his early days in medicine when patients in pain or the late stages of cancer were given orders for heavy medication as a way to assist their "softer exit from this world."[14] The practice was not discussed or debated openly, but was accepted as essential

---

[13] See further responses in ch. 5 on a right to die.

[14] See Francis D. Moore, *A Miracle and a Privilege: Recounting a Half-century of Surgical Advance* (Washington, DC: Joseph Henry Press, 1995) 46. Page references are to ch. 32, "Prolonging Life, Permitting Life to End," published in *Harvard Magazine* (July/August 1995): 46–50.

and non-controversial. Moore believes that assisting a patient to die is vital "to the doctor's historic mission of care and caring for human life, its quality, as well as its duration."[15]

Emerging Technology. A complicating factor is that the present uses of technology pose dilemmas never confronted in a less complicated past. My own growing up was in an era of simple medicine humanely and compassionately administered in a context of small towns and close-knit rural communities. Decisions—even to hasten death—were simpler then because relationships were both more personal and more profound, and the technological options were both far simpler and less extensively available. Whether it was an anomalous newborn or an excruciating and prolonged dying process, the physician could and often did act to end suffering by inviting death. A rural ethos seems to have accepted and confronted death more readily than does a more frenzied and anxious urban ethos.

That approach to humane medicine was greatly modified by the rapid growth and availability of sophisticated technological apparatus available to physicians that enabled them to prolong the life of the dying. Interventions have become so sophisticated that the line between being alive and being dead is neither objective nor obvious. The mandate to "preserve life" might have been clearer and less ambiguous in the era *before* life could be preserved almost indefinitely. Doctors said that Nancy Cruzan could have been maintained with nutrition/hydration for another three or four decades! That was in spite of the fact that she was in persistent vegetative state, from which there has been no documented case of recovery. Preserving life *like that* or *on those terms* seems hardly a moral use of technology. The patient as person has long since disappeared under the myriad hoses, machines, and gadgets that manipulate the vital signs without making an autonomous or meaningful living possible to the person.

Behind the concern for physician-assisted suicide is a profound and thoughtful ambivalence toward life-sustaining medical technology. Yes, patients want it and need it, but patients do not *always* want it or need it. What is feared is that no moral—or even commonsense medical—decision is being made about using technology at the end of life. At times it seems that only a reflex reaction is at work that reaches for a machine rather than deal with the human dimensions of dying. If medicine is to claim to be moral at all, more must be done than simply applying technology as the ever-present "fix" to keep a person from dying. There are times when good ethics

---

[15] Ibid., 47.

requires refusing to use the machine and to allow death mercifully to end bodily existence.

Drs. Quill, Moore, and others recognize that medical technology is better at keeping alive than it is at curing. They are attempting to take accountability for the terribly negative outcomes of applying a technological fix to patient conditions that cannot be cured. Such technology can only prolong the dying process or sustain the patient in that purgatory between life and death where patients are neither clinically dead nor fully alive. When the patient's capacity for willing or intending anything at all has passed, the medical obligation shifts inevitably from keeping alive to facilitating a more comfortable death. Such assistance may take the form either of allowing a natural death or assisting the patient's death.

The new paradigm for medicine is at once more humane and more complex than an ethic built upon an absolute distinction between killing and allowing to die, so strongly argued by Pellegrino and others. First, it recognizes the limits of medicine instead of perpetuating the illusion that medicine is capable of curing all medical ills or of making people immortal. Human medicine is limited—it is a finite enterprise within limits set by human mortality. Medicine does not fail when the patient dies, nor is the physician a failure when a patient is lost to the ravages of disease or injury.

Science has made enormous advances in conquering disease and extending human life expectancy, but people are still mortal; human existence is bounded by birth and death. One illness or another or the ravages of aging will sooner or later overtake mortal existence. Mortality is still the most prominent feature of human existence. Science has no mandate to attempt to make human beings immortal, but it does have a moral mandate to recognize and appropriate the truth of human finitude in dealing with technologies that may prevent or delay death. More accurately, the technologies may provide a rationale for refusing to accept the fact that the patient should be allowed to die.

Toward a New Physician-Patient Covenant

The new paradigm in medicine also recognizes that the patient-physician covenant involves a more complex obligation on the part of physicians than one of simply "keeping alive." The moral obligation includes assisting the patient in his or her dying in appropriate ways. There is a point at which there are no further interventions that promise recovery, but the patient is overwhelmed by pain and suffering. At that point, the patient may decide that death is preferable to a miserable and prolonged process of dying and request assistance from the physician.

The covenant with the physician may involve little more than providing comfort care, using the pain-relieving-and-controlling drugs at the physician's disposal to make one able to live without dysfunctional pain, but it may also involve "ending the suffering (of the patient) as humanely as possible," as Sherwin Nuland says.[16] Two prerequisites would typically be involved in moving to this action: (1) health cannot be restored or the pain cannot be managed; and (2) the patient has requested assistance in dying. The action would be a merciful response to a patient's request, not imposed by a physician because of frustration or malice toward the patient.

These are factors involved in the cases cited by Quill, Nuland, Moore, and countless others. One surgeon tells of treating a man in his seventies who had reached the terminal stages of leukemia. Despite antibiotics and surgery, an abdominal abscess developed that was impossible to manage despite repeated surgeries and interventions. The man was now subject to dying slowly and with intractable pain during which he had lost all semblance of dignity. The surgeon's brother-in-law physician suggested he be assisted to die more quickly, with which the surgeon agreed. He injected a lethal dose of morphine into the patient's intravenous tubing.[17]

A perception that medical technology is without controls and insensitive to the deepest human needs in dying is a major factor supporting the new medical ethic. The image of medicine at its beneficent best is difficult to maintain in light of the images of dysthanasia, or bad dying, now indelibly etched in the modern mind. It is the image of William Bartling, afflicted with five lethal ailments, struggling against and declaring he wanted the tubes withdrawn while being tethered to his bed to prevent self-extubation. Another is the image of the Alzheimer's patient who had lost all ability to communicate, care for herself, or walk. She could only scream *feuer!* (fire!) in her native German. Whether she was feeling on fire with an unquenchable heat or terrified at seeing a raging flame was impossible to tell, but out of his desperation and love for his wife of many years, her husband put an end to her suffering with a pistol shot to her head.

Ironically, medical technology developed as a promise to increase our freedoms and expand our options for living. Good medicine is promoted by the hope for cures and longevity, but the technology of life support seems to

[16] Sherwin B. Nuland, *How We Die: Reflections on Life's Final Chapter* (New York: Alfred A. Knopf, 1994), 5.

[17] Ibid.

have become an end in itself based on the problematic notion that "what can be done should be done." Technology is now a part of what Verhey calls the determinate features of our existence.[18] The medical mandate seems to be to use life-sustaining equipment *regardless* of the medical condition of the patient or the absence of a reasonable promise of patient recovery. Medical technology introduced and implemented as an option to employ when medically appropriate seems now to have become an inescapable obligation.

Most all the dramatic actions taken to end the suffering of loved ones in recent years from withdrawing their "life" support systems to merciful killings (suicide, homicide) are signs of the desperation felt by people when confronted by irreversible dying, unbearable suffering, and seemingly inescapable technology.

Surgeon Dan English once called some of the treatments imposed on patients in the ICU the "crucifixion of the dying." As he pointed out, the ethical question in medicine is not just what is done *for* people, but what is done *to* people. The unrelenting application of unfeeling but sophisticated technology in the name of patient care and the morality of refusing to cooperate with death is suspect at best, immoral at worst. Medicine is facing a crisis of the morality of aggressive care in the use of expensive, uncaring technology that is applied at the end of life. The machine does its job whether or not the patient has long since passed the point of being restored to meaningful life.

Medicine has come a long way in its ability to prolong our lives and often reverse the effects of a traumatic health crisis, but it is not always able to cure. Deep in the human psyche now lies a new and fearful image—that of dying like Karen Anne Quinlan, Nancy Cruzan, or Terri Schiavo (after ten to fifteen years in persistent vegetative state while maintained with gastronomic tubes for nutrition and hydration), or of a William Bartling, dying of five lethal and incurable diseases, insisting on withdrawal of the tubes and wires, but constrained by tethers on his arms so he could not disconnect himself.

The Patient Self-determination Act, passed by Congress in 1989, was a legislative response to the enormous pressures from the public for relief from "paternalistic" medicine from which patients could hardly escape. Only the most addicted medical technocrat believes it is ethical to use technology aggressively when it is unable to restore health or violates the wishes of the patient.

---

[18] Verhey, "Choosing Death," 718.

Accountability for Bad Outcomes. Another element in the new covenant between physician and patient is that accountability must be required for bad medical outcomes when technology fails to restore health. Who will be responsible when the patient can be weaned from the ventilator, but cannot and will never think or know anything or anyone at all? Who will be responsible for the fact that the patient was in fact clinically dead, but resuscitated in order to see if health could be restored, then wound up in a condition of perpetual dependency on heartless technology? Who will take responsibility for the severe neurological impairments with which many live who were "saved" by medical technology when they entered the world as terribly premature or low-birth weight neonates? They had been born dying, but death was delayed by the artful uses of sophisticated technology, only to sentence the child to a life of severe impairment.

The story of Brenda Young of Flint, Michigan, captures the tragedy of technology without a transcending ethic. Since a seizure in 1992, Young had needed total care; she had to be fed, bathed, diapered, and tied to her bed at night lest she push herself over the padded bedrails. Her anguish was communicated through unintelligible screams; the few words comprehensible were "water" and "bury me." She had tried to avoid this kind of existence by signing an advance directive before the incapacitating seizure. She also gave her mother power of attorney and directions to stop treatment if she became incapacitated. All that was to no avail; she was placed on a ventilator, tube-fed, and maintained through a two-month coma, despite her mother's insistence that such support not be used.

Young and her mother have now won a $16.5 million lawsuit against the hospital, Genesys St. Joseph. They may represent a new wave of lawsuits seeking to hold hospitals, nursing homes, and physicians liable for ignoring the mandate of advance directives and/or medical surrogates. The charge was that the imposition of unwanted, unapproved, invasive medical treatments constitutes a type of battery, as the Supreme Court said in *Cruzan*.[19]

Frustration with Punitive and Outdated Laws. A further factor behind the public's push for laws allowing physician-assisted suicide is the reprehensibility of punishing doctors for a merciful and morally justifiable action. Arguing that it is not an immoral act holds that it is prompted by beneficence, not malice; thus, the practice should not be regarded as unethical by professional standards nor criminal by legal standards.

---

[19] The story was carried by *The Courier-Journal*, 2 June 1996, A11.

Surveys of physician practice make the point that actions like those of Nuland, Quill, and others represent in public what is somewhat commonly practiced in private. When asked anonymously to indicate whether they had ever assisted a patient's dying, many physicians and nurses will admit to having done so.

A national study of critical care nurses found that nineteen percent had engaged in euthanasia or had hastened a patient's death by only pretending to provide life-sustaining treatment ordered by a physician.[20] Seventeen percent of respondents had been requested to provide euthanasia or assist a suicide, and nineteen percent of that group reported that they had actually done so.

A study of physician-assisted suicide and euthanasia in Washington found that twenty-six percent (218 of 828) of physicians had been asked by patients to help them die. Prescriptions for a lethal dose of medication were provided for 38 of 156 patients. Fifteen of the patients did not use the medication.[21]

Dr. Nuland tells of a leading oncologist who had kept count of the patients who had asked him to help them die. "There were," he said, "127 men and women, and I saw to it that 25 of them got their wish." Nuland admits he "has done exactly as he has," except "on far fewer occasions."[22] The decision, he says, was done in the sanctity of the privileged doctor-patient partnership. "We have done it for people we know well," he says, "whose desperation for the relief only death can bring seemed entirely appropriate."[23]

The actions range from assisted suicide, to withdrawal of treatment, to euthanasia. Neither the decision nor the action taken was motivated by malice, impatience, or mercenary considerations. They were thoughtful, considered, and compassionate responses to the devastating circumstances of the patient and the imminence of death. The patient was overwhelmed by disease or suffering. It was time to let them go.

Had such caring physicians been subjected to criminal penalties, the tragedy of an ugly death would have been compounded by a miscarriage of justice. The tragic loss of Diane to leukemia would have been sadly multiplied had her husband, son, and/or Dr. Quill been indicted for assisting her death because of knowing her intentions without reporting or

---

[20] *Medical Ethics Advisor* (*MEA*) (July 1996): 74.

[21] Ibid., 82.

[22] See his interview in *USA Weekend*, 3–5 February 1995, 4–6.

[23] Ibid., 4.

intervening to stop her action. The fact that Quill and other physicians who had signed on as *amicii* were not arrested by Attorney General Vacco is a sign of the flexibility of the law when there are mitigating circumstances involved in an activity that otherwise might be regarded as criminal.

Oregon's Measure 16 is an effort to humanize the system where the law is concerned. The people of Oregon seem to believe that a compassionate physician who assists the dying is a friend, not a foe or a fiend in disguise. Such physicians attempt to practice both good medicine and good morals. The image of family and friends gathered around Dr. Kevorkian with hugs and congratulations at his acquittal in Michigan was a telling sign of the relief and joy when merciful actions transcend inadequate and unjustifiable laws. If we can imagine that death under certain circumstances is desirable and that assisting one to escape a miserable dying is an act of mercy and love, we can also believe that the law should recognize the absence of malice in that assistance. There is no justification for physicians having to live in fear of exposure to public ridicule or censure or to legal prosecution when they have acted in good faith to assist one's dying a better death. [24]

I sympathize deeply with the anguish and shame felt by Dr. Quill in having to abandon Diane in her dying moments. He denied the ethics of compassion and the ordeal of moral medicine because of fear for his own professional well-being. Unfortunately, doing what was right both medically and morally placed him in jeopardy with the law.

Personal Fears of Dependency and Loss of Control. The call for physician-assisted suicide also reflects concerns about human dignity in dying. The image of the person is that of a creature who is to a degree responsible for "setting the terms" with death. Dr. Susan Goold, of the University of Michigan, Ann Arbor, has said that "public support of physician-assisted suicide is a reaction to fear among patients that they will lose their rights when they become terminally ill."[25]

A study in Washington led by Anthony L. Back, geriatrician and member of the ethics committee at Veterans Affairs Puget Sound Medical Center in Seattle, found that "powerful things...affect people's thoughts about suicide, including many non-physical things...(which) physicians have never thought about and have no idea how to address."[26] The most powerful thing, he said, was the patient's need for a sense of control. Physician

[24] See Timothy Quill and Diane Meier, "The Big Chill—Inserting the DEA into End-of-life Care," *NEJM* 354/1 (January 5, 2006): 1–3.

[25] *Medical Ethics Advisor* 11/2 (February 1995): 13.

[26] *MEA*, July 1996, 74.

response to requests for assisted dying was typically to treat physical symptoms. Even referrals for psychological consult happened in only twenty-four percent of cases (50 of 204) involving requests.[27]

A Christian theology of personhood might also support the concern for dignity in dying. Christian anthropology begins with the notion that persons are created *in the image of God*, which means (among other things) that people know in unique ways. Of all creatures on earth, only people know they are to die, can anticipate its coming, and make plans accordingly based on their own framework of beliefs, commitments, and responsibilities. Human dignity consists in this process of reflective anticipation and introspective preparation for this ultimate venture in life.

People deny and are denied human dignity when they remain the passive, helpless victim of dysfunctional or meaningless suffering or the ravages of non-reversible disease that wastes the body and devastates the mind. By definition, these processes erode the very powers that make human dignity possible. Sigmund Freud had lost any semblance of the dignity he had achieved as a man of extraordinary ability and insight who made a remarkable contribution to the world of psychiatry. To waste away without any control of his bladder or bowels was an unbearable indignity for him.

Dr. Quill says he supports "a patient's right to die with as much control and dignity as possible." Christian moral commitments do not require that one be the purely passive victim of a horribly ravaging and unrelenting disease. One is not morally obliged to accept the physical debilitation and mental deterioration that slow death inevitably brings. Stoic passivity is hardly moral heroism.

Dan Brock makes the point in terms of personal autonomy. "By self-determination as it bears on euthanasia," he says, "I mean people's interest in making important decisions about their lives for themselves according to their own values or conceptions of a good life, and in being left free to act on those decisions. Self-determination is valuable because it permits people to form and live in accordance with their own conception of a good life, at least within the bounds of justice and consistent with others doing so as well."

Brock regards self-determination as a central aspect of human dignity. The dying person may therefore decide "that the best life possible...(even)

[27] Ibid., 82.

with treatment is of sufficiently poor quality that it is worse than no further life at all."[28]

A little-explored aspect of this openness of the future with regard to one's own death is the sense of freedom it provides many people as they face the inevitable. Nancy Dorr, who has breast cancer, supported Oregon's "death with dignity" act. When asked why, she reported that her tumor had grown twenty to thirty percent since her last doctor's appointment. The passage of the legislation gave her a sense of relief to know she would not have to suffer long if her illness became terminal: "You can't believe how freeing that is. It gave people with terminal illnesses control over their lives. When the quality of life isn't there, [you] should have the option not to suffer."

The Desire to Have Someone Present to Us in Our Dying. The late Paul Ramsey coined the term *agathanasia* (*agathon*, good; *thanatos*, death) as an alternative way of viewing what is sometimes called "euthanasia." A central contention in his argument was that it is immoral to isolate and abandon the dying person. His insight was that there are few things more important to us when in pain or facing the final throes of dying than the presence of family and loved ones. [29] Few people want to die alone. We want those significant others in our lives to be present to and with us, to touch us, to talk to us, to comfort and counsel us.

A major part of the tragedy in Diane's death was the fact that she died alone. Neither her husband, her adult son, nor Dr. Quill felt free to be with her, to talk to her, touch her, or comfort her in her dying moments. Had they remained with her, they could have been vulnerable to prosecution under New York's law against assisted suicide, but she would have had the company and comfort of people important to her, as she was to them, in her final moments of living.

The law and its interpreters are not always able to deal with moral nuance. Subtle but all-important distinctions are certainly required when discerning the difference between morally licit or morally illicit killing. Harsh and punitive laws drive loved ones and family away from the dying when a choice is made to die. If care has been taken to assure patient preferences, and the medical prognosis of the imminence of death is

---

[28] Dan W. Brock, "Voluntary Active Euthanasia," *HCR* 22/2 (March/April 1992): 11.

[29] Paul Ramsey, *The Patient as Person* (New Haven: Yale University Press, 1970) 72.

supported by specialists who have been consulted, the decision should hardly be regarded as either morally or legally culpable.

Hospice care is structured around the medical concern for patient comfort. Every effort is made to comfort patients during their dying. The covenant with caregivers is that aggressive interventions will not be imposed, but the pain will be managed, even if that might result in shortening the dying process.

Dying in a Community of Caring and Moral Support. Apparently, there is widespread moral support for aiding a freely chosen path toward dying when death is inevitable and imminent. There is even moral support for suicide when life is unbearable, as from intractable pain, whether psychic or physical. Some of Kevorkian's patients were people who made choices that involved quality-of-life considerations; they were not all terminally ill.

The support for physician-assisted suicide is based on the idea that it is moral and therefore medically acceptable *under certain circumstances*. When a patient cannot be returned to meaningful living, reasonable efforts have been exhausted, and the patient is overwhelmed by pain, death may be not only inevitable, but desirable. Those who argue that life is "sacred," *regardless* of its circumstance, are either not addressing the horrible ways in which people may and do die, or they are expressing personal fears about death, but to claim further that choosing death is immoral and religiously unsupportable is to make a value judgment that is neither universally shared nor consistent with biblical perspectives. King Saul *chose* to die either at his own hand or that of his armor-bearer (cf. 1 Sam 31:1–5; 2 Sam 1:1–16), and the Apostle Paul indicated that he had not yet *chosen* whether to remain alive or go on to be with the Lord, which he regarded as much more desirable than remaining alive under his present circumstances (Phil 1:23).

Paul was apparently stating openly an option he felt existed for him to carry out privately. Public openness is an important principle of moral action. Private acts have community dimensions. People should not be forced to act alone or in isolation. As Nuland puts it, "What troubles me is the very privateness of my decisions." He was not bothered about its morality, but its covertness. He wanted a way to make the decision more open, more subject to a broader-based decision-making process, so as to protect it from abuse. He thinks neither the Netherlands nor the Oregon law is adequate. Nor is it sufficient to rely upon the personal morality of the physician in context of "a long and empathetic relationship."[30]

---

[30] *USA Weekend*, February 3–5, 1995, 4.

Just how he would implement the controls he has in mind is more difficult to discern. Cogent objections to assisted suicide are not difficult to make, but it is nearly impossible to suggest an alternative that is morally superior.

Diane was supported in her decision about dying by a moral community that surrounded her with care and personal presence—at least until the very end. Such support does not remove, but it at least mitigates, the tragedy of an untimely death. The morality of suicide is at best ambiguous. Many people believe that, under certain circumstances, it may be a positive good. There are things worse than death, and dying slowly with pain from an ugly, debilitating, and depersonalizing disease is one of them. We need not settle the question of the morality of suicide to be able to say very clearly that the dying should not be abandoned to die alone.

Nuland called for a review panel composed of physicians, specialists, etc., as a way to safeguard "a community of support" for such decisions. He called it a "council of sages."[31] Such "a group of people representing society's collective wisdom" would likely be both unmanageable and unworkable. How could such a group function openly in a society already so deeply divided along ideological and religious lines that abortion providers are threatened with death and Terri Schiavo became the subject of political action groups and public demonstrations that reached to the White House? America has political and moral gridlock in discussions of some of the most vital issues in medical ethics. "True believers" dominate all such discussions with an attitude that destroys both civility and the possibility of reaching consensus on complex questions.

Instead of such a panel, Quill suggests something like the rules that govern the process in the Netherlands. It is sufficiently structured so as to avoid hasty and ill-advised, unilateral decisions on the part of the physician and/or patient and family, but sufficiently protected from the public purview as to make decisions possible that could be called "moral" in a medical context where death and dying must have their recognized place alongside life and health.[32]

The Acceptance of Death. A final but major factor in support for assisted dying is the realization that there are many things worse than death, including certain circumstances of living. "It is not death," said Moilière,

---

[31] Ibid., 6.

[32] Timothy Quill, "Doctor, I want to die. Will you help me?" *JAMA* 270/7 (August 18, 1993): 872.

"but dying that worries me."[33] From a Christian perspective, death is neither an absolute enemy nor an ultimate threat to human well-being. As Allen Verhey puts it, life and its flourishing are not ultimate goods. Quoting Karl Barth, he argued that life is not to become a "second god." Nor are suffering and death ultimate evils.[34]

The late James Weldon Johnson captures this Christian belief in his great poem-sermon, titled "Go Down, Death." He portrays "Sister Caroline," down in Yamacraw, Georgia,

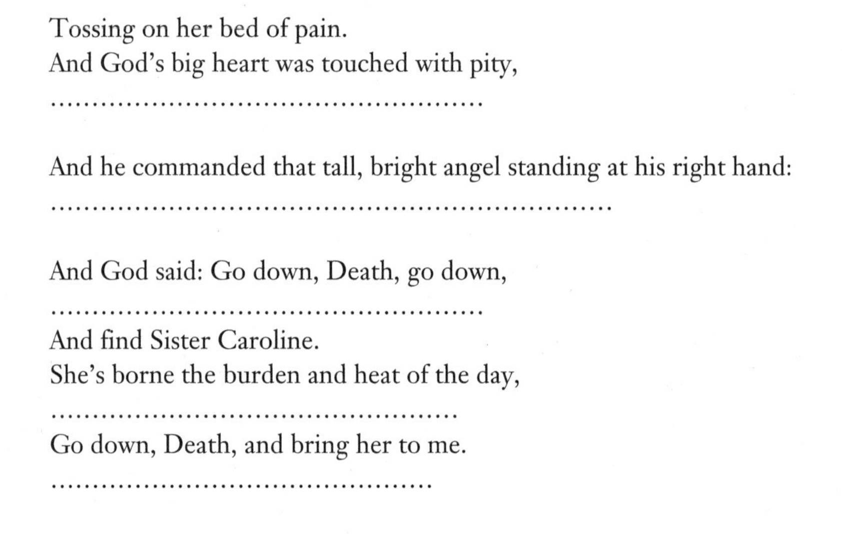

> Tossing on her bed of pain.
> And God's big heart was touched with pity,
> ....
>
> And he commanded that tall, bright angel standing at his right hand:
> ....
>
> And God said: Go down, Death, go down,
> ....
> And find Sister Caroline.
> She's borne the burden and heat of the day,
> ....
> Go down, Death, and bring her to me.
> ....
>
> While we were watching round her bed,
> She turned her eyes and looked away,
> She saw what we couldn't see;
> She saw Old Death....
> Coming like a falling star.
> But Death didn't frighten Sister Caroline;
> He looked to her like a welcome friend.
> And she whispered to us: I'm going home,
> And she smiled and closed her eyes.

As Johnson so poignantly portrays it, death may be welcomed as an intimate friend and deliverer from the evils of suffering and the ravages of

[33] Molière, "The Empty Chair," in Martin Crowley, ed., *Dying Words: The Last Moments of Writers and Philosophers* (Amsterdam: Rodopi Press, 2000) 25.

[34] Verhey, "Choosing Death," 717.

pain in the process of an ugly and prolonged dying. One woman prayed for the death of her husband whose anguish from cancer had become unbearable. His days and nights were spent in unrelieved agony. Life had neither value nor hope; death was everywhere present except in reality. When he finally breathed his last, she said, a blessed wave of peace swept over his face. And she bowed her head and thanked God for the good gift of merciful death for her beloved husband.

This more ready acceptance of death might at least in part be attributed to the success of death education seminars during the past three decades. The studies of Elizabeth Kubler-Ross and others have contributed to a more enlightened approach to the subject. Death is being demythologized.

Ernst Becker, in his Pulitzer Prize-winning book *The Denial of Death*,[35] pointed to the power over the human psyche held by the fear of death. Most human actions, he said, were determined by the refusal to come to terms with the fact of human mortality—to accept it not only as inevitable, but in the last analysis, desirable. Among the denial syndromes he cited were those desperate efforts to resuscitate the hopelessly dying patient and the misguided efforts to discover the elixir of immortality.

The notion of "the sanctity of life" loses any semblance of truth-value when "life" has been reduced to the last vestiges of the vital signs or the final struggles of our bodies with the diseases that overwhelm us. The Apostle Paul had it right in saying that "if for this life only we have hope, we are of all people most miserable" (1 Cor 15:19).

There is widespread interest in something like the *ars moriendi* (the art of dying), which proliferated in the Middle Ages. In these exercises, the dying prepare themselves spiritually for "giving up" their spirits to the Eternal Spirit. The aim was to die without fear or struggle against the inevitable—to die with equanimity, courageously facing death rather than "striving officiously to keep alive" or struggling desperately against the coming of death.

Making decisions to hasten death also seems consistent with a Christian theology of death and resurrection. Death is not the final victor for those who die in faith. A Christian approach to death maintains that the power of death has been overcome in Christ's resurrection. It is not death, but life, that has the final word. As Paul put it, "Death is swallowed up in victory" (1 Cor 15:54d). Christians can say confidently with the great

---

[35] Ernst Becker, *The Denial of Death* (New York: Free Press, 1973).

Apostle that "whether we live or whether we die, we are the Lord's" (Rom 14:8).

The biblical revelation provides two reasons for viewing death as a good gift. First, it may alleviate an unbearable pain, as in the cases of Samson (Judg 16:28–30) and Saul (1 Sam 31:4–5). Second, it is placed in a positive light as a transition to eternity. "What you sow," said Paul, "does not come to life unless it dies" (1 Cor 15:36). "It is sown a physical body, it is raised a spiritual body.... For this perishable nature must put on the imperishable, and this mortal nature must put on immortality" (vv. 44, 53).

It is this ground-of-meaning belief that brought Paul to see death as a way to escape his tribulation and be with the Lord (Phil 1:23). He was, he said, torn between two opinions or options: "to be with the Lord, or to remain with you.... Which of these *I shall choose* I do not yet know" (v. 22b, emphasis added), he declared, apparently indicating the choice was his to make. He preferred, he said, "to depart and be with Christ, for that is far better" (v. 23), but commitments to and sense of responsibility for the Philippians made him willing "to remain in the flesh" (v. 24).

Conclusion

Rapid developments in medical technology have brought about a new era in our relation to death and dying. The ethics of (only) prolonging life by the aggressive use of technological interventions or to leave dying totally "in the hands of God" are being challenged as inadequate at best, irresponsible at worst. Strong voices are saying that such approaches do not adequately account for human moral responsibility in the medical context.

Physicians like Quill and Nuland have come to embrace and embody a new and more burdensome dimension of caring for the dying. Loving care goes beyond the ethic of simply "keeping alive" far past the time of meaningful existence. Assisted dying may also be the requirements of love and mercy when the patient is overwhelmed by disease. The covenant between physician and patient might well include aiding a quicker and less painful death. Morality requires that the patient not be abandoned, but be comforted, consoled, counseled, and cared for even if they choose death.

Those who choose suicide rather than await the final ravages of debilitating illness may well be expressing the uniquely human capacities for dignity even in the face of defeat. Such decisions may signify a commitment to the truth that "whether we live or whether we die, we are the Lord's."

And those physicians who assist the dying to die more quickly in the face of an ugly, prolonged bout with unbearable suffering may well be embracing a profound but very difficult moral responsibility. They do not run from the burden and threat involved in what Paul Ramsey called "the countervailing requirements of agape." [36] They stand with their patients and do not abandon them to the isolation of solitude as they confront eternity. Such physicians embrace neither a life worship that is blind to life's quality, nor to the claim laid upon them by the patient's request.

The future of physician-assisted suicide is thus rather uncertain. Likely, the practice now rooted in the public mind will not go away. The moral revulsion expressed and the public sentiment against PAS are not sufficiently grounded in ethics and theology to offset the support generated for the practice among people who see the issue in profoundly different ways. The practice of physician-assisted suicide will undoubtedly persist. Doctors and nurses will continue to help the dying in ways mutually agreed upon and in a manner mutually acceptable, regardless of the legal statutes of various states. PAS has been and likely will remain a part of medical practice as long as there are people who want to die with a caring physician's assistance. They may well place their decision in the context of a religious faith that has personal depth, biblical warrants, and widespread public support.

---

[36] Paul Ramsey, *The Just War: Force and Political Responsibility* (New York: Charles Scribner's Sons, 1968) 206.

Chapter 7

ETHICS AND THE ARTIFICIAL HEART: CYBORGS AND THE HUMAN FUTURE

In the Old Kingdom there was a pharaoh named Khufu who summoned the wizard Djedi to his court. He asked of Djedi, "is it true, what they say, that you can join a severed head?"

"Yes, I can, O King, my lord," replied Djedi. And a goose was brought to him and its head cut off. And the head was placed on one side of the great hall and the goose was placed on the other. Djedi had his say of magic and the goose began waddling, and its head did also, and when one reached the other they joined and the goose stood cackling. Then a long-legged bird was brought, then an ox, and the same was done to them. And in each case, Djedi joined their heads.

Now King Khufu ordered that a man be brought from the prison so Djedi could use his magic on him.

Cried Djedi: "But not to a human being, O king, my lord! Surely it is not permitted to do such a thing to a man."

—From the Papyrus Westcar, "Three Tales of Wonder"[1]

Djedi had it right. There are some things that ought not be done to human beings. Widespread agreement on this point can be found among professionals in medicine, law, religion, and philosophy, but there are considerable differences of opinion as to what things ought not be done. The Djedi story assumes that the very nature of a particular action ruled it out for use on a person. Severing a head—even if it could be reattached and that without harm to the person—simply should not be done.

---

[1] "Three Tales of Wonder, " from the Papyrus Westcar, in J. B. Hare, ed., Internet Sacred Text Archive, Scriptures of All World Religions, www.sacred-texts.com/index.htm.

Might the same be said about replacing a human heart with one that is totally artificial? Or what if a totally artificial person, a cyborg, could be developed? Should that be done? Should the effort to design and manufacture a robot made to look and function much like a human being be condemned by good morals, rejected by scientists, and prohibited by law?

The bioengineers are hard at work. While philosophers and theologians debate the ethical issues, technological procedures continue without interruption. There seems no area of human life that is off-limits for experimentation or modification as long as the rationale of providing extended life can be given. Experiments with a totally artificial heart (TAH) crossed an important line in the psychology of science. The mystique of the heart and its ties to what it means to be human have been challenged and found inadequate as a bulwark against invasive and ever-expanding biotechnology.

The Artificial Heart

The artificial heart is now receiving worldwide attention. Each device is well publicized and widely touted as offering hope for extended life and inestimable commercial value. Some are proposed as permanent replacements of the heart, others as assist devices, some as bridge-to-transplant supports. Plastic and titanium, wires and batteries, they all seem indispensable in the world of thoracic surgery. The engineer and the surgeon are partners in medical science.

Some mechanical hearts have attained therapeutic status; most are approved only as experimental devices. Some are pulsatile, mimicking the beat of the heart; others are steady stream devices that leave the patient without a discernable pulse. They range in size from the mighty Jarvik 7 to the OL 80, an electric heart-assist device that is no larger than a bicycle bell. It weighs eight ounces and has one moving part, a six-bladed turbine driven by electromagnets. The Jarvik 2000 may be the smallest of all. It weighs in at ninety grams and is roughly the size of a human thumb.[2] Miniaturized devices help address an imbalance in the distribution of life-prolonging devices. People with smaller chest cavities, such as women and the Japanese, were excluded as recipients of such hearts as Jarvik 7[3] and the more recent AbioCor.

---

[2] Go to www.texasheartinstitute.org. Follow links to Research and Jarvik 2000.

[3] No women were included in the Jarvik 7 trials because of their generally smaller thoracic cavity. Jack Burcham died after ten days because his chest could not be

Ventricular assist devices (VAD) offer another way to deal with a failing or inadequate heart. Instead of displacing the heart, they are attached to or installed within it. Ventracor of Australia has the VentrAssist, which is about the size of a tennis ball. CardioWest, a direct descendent of Jarvik 7,[4] and Thoratec are pulsatile, as is the HeartMate, which is being developed currently at the Texas Heart Institute.[5] Japan is working with the tiny AB–180, a heart-assist device that is driven by electromagnets. [6]

Such devices have their own attraction as medical interventions for dying patients,[7] but the TAH poses ethical issues that have their own troublesome persistence, if not uniqueness. Difficult questions are posed, ranging from the process to obtain informed consent; questions about the artificial versus the natural; social demographics and burgeoning healthcare costs; access to exotic treatments; and the place of death when making policy decisions about whether indefinitely to extend the life of elderly, dying patients.[8]

At present, the aim behind the use of the artificial heart is ostensibly to prolong the life of the patient or improve his or her quality of life. A more global question is whether the artificial heart is the beginning of an era in which the human is being displaced by the artificial. Experiments are proceeding in many areas as integral parts of the effort to deal with human mortality and the breakdown of organs vital to human functioning.

The visually impaired are offered the hope of recovering vision through computer and video technology. Those who cannot hear are given implants that enable discernment of patterns of voices and thus more personal communications. Sexual identity can be clarified by surgery and enhanced by chemistry. Dialysis has been available for years to

---

closed. See R. C. Fox and J. P Swazey, *Spare Parts: Organ Replacement in American Society* (New York: Oxford University Press, 1992) 132.

[4] J. G. Copeland et al., "The CardioWest Total Artificial Heart Bridge to Transplantation: 1993–1996 National Trial," *Annals of Thoracic Surgery*, vol. 66 (1998) 1662–69. See also J. G. Copeland, "Current Status and Future Directions for a Total Artificial Heart with a Past," *Artificial Organs* 22/11 (1998): 998–1001.

[5] D. A. Cooley, "Mechanical Circulatory Support Systems: Past, Present, and Future," *Annals of Thoracic Surgery*, vol. 68 (1999) 641–42.

[6] N. Honda, et al, "Ultracompact, Completely Implantable Permanent Use Electromechanical Ventricular Assist Device and Total Artificial Heart," *Artificial Organs* 23/3 (1999): 253–61.

[7] See Fox and Swazey, *Spare Parts*, 11–12.

[8] See Daniel Callahan, *Setting Limits: Medical Goals in an Aging Society* (New York: Simon & Schuster, 1987) 53. See also Callahan's, "Rationing medical progress: the way to affordable health care," *NEJM* 322 (21 June 1990): 810f.

accommodate failed kidneys, and lungs are assisted by ventilators. Prosthetic devices are both functional and realistic-looking. Skin and retinas can be developed artificially. In short, the future holds the possibility of the totally artificial body—a robot with all the emotional and intellectual features we associate with being human.

The symbolic and ethical significance of such giant steps in medicine is of tremendous importance. People disagree as to whether we are making progress or simply making history. At a minimum, however, the effort to develop artificial organs represents a significant step in the effort to conquer disease and cope with human mortality. Such devices are a step toward realizing the hope for greater control over death and life-threatening events.

The consensus among scientists and others is that an artificial heart will be developed and that it will be therapeutic, reliable, and widely accepted.[9] There is far less agreement, however, as to its meaning for the human future.

The Heart, Cyborgs, and the Human Future

There is little question that the artificial heart will have applications far wider than for patients with end-stage heart disease. Scientists and futurists are expecting to develop totally artificial persons, or cyborgs, of which the heart is only one component. A cyborg is a robot that has features of intelligence and personality modeled on or derived from people. Thoughtful people in science and the humanities are debating whether cyborgs would be human since they have characteristics typically associated with human beings. To put it differently, to what extent would a person's body have to be modified by technological—that is, artificial—components for them to become a cyborg?

People have become the object not just of experimentation, but of modification by design and purpose. Science fiction may be a near-realistic portrait of the world to come. Fascinating stories of the totally bionic person have already appeared, in which the technological is combined perfectly with the biological. *The 6 Million Dollar Man* and *The Bionic Woman*, not to mention the more recent *AI: Artificial Intelligence* and *I, Robot*, portray the technologically sophisticated displacing the biologically limited, damaged, or degenerated. Replacement parts are constantly and readily available in

[9] The FDA has now given approval for AbioMed to sell up to 4,000 devices per year. "FDA approves AbioCor heart," *The Courier-Journal*, 6 September 2006, A1, 11.

that world of fantasy becoming reality. We seem to be moving toward or becoming a world of cyborgs.

Cyborgs and the Future

The genetic/bioengineering revolution is proceeding rapidly. The "engineering" model, which combines technology with the creative capacity for designing bodily replacement parts on a wide scale, now dominates science. Genetics, nanotechnology, and robotics are basic to the scientific revolution and the future of the new being.

The new knowledge of genetic malleability marks a turning point in the evolution of life, says Robert Sinsheimer. People are the first creature both to understand its origins and undertake to design its future.[10] Recombinant DNA makes it possible to switch characteristics belonging to one entity to another and thus transform the recipient. Thomas Eisner says bluntly that a biological species should "be viewed as...a depository of genes that are potentially transferable. A species is not merely a hard-bound volume of the library of nature. It is also a loose-leaf notebook, whose individual pages, the genes, might be available for selective transfer and modification of other species."[11]

People can also be modified by wedding technology to human biological material. The DNA computer chip is already on the drawing board and signals a technology that will bring about a revolutionary future.[12] Nuclear technology, once thought the hope for the future for humanity, has now become suspect and threatening. The hope it embodied is now thought disillusioning because of its threat to life itself, b ut science now poses the possibility of altered human states—new minds and new bodies—which some view as a new hope for the human future.

---

[10] Robert Sinsheimer, "The Prospect of Designed Genetic Change," *Engineering and Science* (April 1969): 8–13.

[11] Thomas Eisner, "Chemical Ecology and Genetic Engineering: The Prospects for Plant Protection and the Need for Plant Habitat Conservation," Symposium on Tropical Biology and Agriculture, St. Louis, Missouri, Monsanto Company, July 15, 1986.

[12] "Computation Takes a Quantum Leap," *Science News* 158 (August 26, 2000): 132. IBM Almaden Research Center in California says a basic model of a quantum computer has already been designed. Five flourine molecules are combined with a few other atoms and the product is both a molecule and a computer. The quantum computer will have many times the power of a conventional computer since a qubit can take on more values simultaneously.

Nobel laureate biologist Joshua Lederberg coined the word "algeny" to capture this notion of the transformability of one organism into another. The term comes from "alchemy," which attempted to enhance the process by which it was believed that all metals were becoming gold. Alchemists were committed to accelerating this natural process toward the perfection of or the intended end of ordinary metals. It became a mystical and spiritual, not to mention a lucrative commercial, venture.[13] Similar motives and dynamics can be found in the push for developing cyborgs on a wide scale.

Rodney Brooks, director of the Artificial Intelligence Lab at MIT, says the development of cyborgs is simply a consequence of standard practices among scientists who move from analysis to synthesis, from science to engineering. A cyborg, he says, will be little more than a marriage of silicon and steel with biological matter.[14] What is more, he says, with the rapid proliferation of functional prosthetics we are all becoming bionic.[15]

William Gibson, the writer who coined the term "cyberspace," notes that the future will probably see implantations of silicon chips into the human brain, which have been modified with DNA. Such "inelegant procedures" will nonetheless find a rationale, he predicts, whether by the military or mainstream medicine.[16] Models of nanomolecular computing are already emerging. Scientists can combine brain cells or human DNA with gellatinous computing "goo" and get all the benefits of both artificial and natural intelligence.

If such developments are mastered and the technique perfected, the twenty-first century will be a world in which the differences between the artificial and the natural will be increasingly difficult to discern. Machines will be made in the image of human beings—intelligent, thinking, calculating, feeling—and thus also, like people, concerned with survival, replication, and control of the environment and dominance over competitors.

Brooks says we are making progress. Robots are emerging that have human feelings, wants and needs, reflective thinking, and self-perpetuating capacities. Computer programs reproduce and evolve. They exhibit actions

---

[13] Cited by T. Burchhardt, *Alchemy: Science of the Cosmos, Science of the Soul*, trans. W. Stoddart (London: Stuart and Watkins, 1967) 25.

[14] Rodney Brooks, *Flesh and Machines: How Robots Will Change Us* (New York: Pantheon, 2002) 185.

[15] See B. Bower, "Mind-expanding Machines," *Science News* 164 (August 30, 2003): 136.

[16] W. Gibson, "Will We Plug Chips into Our Brains?" *Time* (June 19, 2000), Supplement.

and patterns we once expected only from living creatures. They interact with complex environments, chase prey, evade enemies, and compete for space and environment.[17]

To be sure, like the artificial heart, today's robots are incomplete and imperfect. Their technological design still lacks the sophistication that will make them an acceptable alternative to human biological existence, but they are evolving, assisted by their curious, inventive and determined creators who are also their masters at this stage of technique history. After all, they will be like people, that is, imperfect. Eventually, says Brooks, we may extend to them certain inalienable rights much like those to which people are now entitled. The more "human" they become, the more limited the options for dealing with them should they become fierce and clever competitors, if not an open and hostile enemy.

The scenario is the stuff of science fiction, but it is not entirely imaginary or fictional. Creative and scary stories have come from those who imagine the future in terms of projected trends of the present. The data with which they work are not entirely fantasy—the facts are there in fragmentary form. All that remains is "an intuitive grasp of wholes"—bringing the parts together into some type of comprehensive story that makes sense of the data and sets forth a predictable (if not altogether reliable) trajectory into the future.

Cyborgs as Salvation: Death the Enemy

Enthusiastic supporters of the new biotechnologies eagerly anticipate the age of cyborgs. The new utopians, or "technopians," believe technology is the key to achieving the goals of medicine more quickly and efficiently. The World Transhumanist Association is committed to the notion that the new world of the future is available by reason, science, and technology. They value respect for the rights of individuals and a belief in the power of human ingenuity. They repudiate any reliance on the existence of supernatural powers that guide nature or the future. As they put it, "The critical and rational approach which transhumanists support is at the service of the desire to improve humankind and humanity in all their facets."[18]

Also called extrapoanists,[19] they seek technopia by merging the human with the machine, which they see as the only way people will be able to

---

[17] Rodney Brooks, "Will Robots Rise Up and Demand Their Rights?" *Time* (June 19, 2000), Supplement.

[18] Go to their website at: www.transhumanism.com.

[19] Brooks, *Flesh*, 208.

survive. Their strategies blend human hopes for ending pain, suffering, and limitation with the prospects for change made possible by scientific innovations.

Downloading consciousness into a computer or robot is thought of as one way to live on when one cannot go on as a mortal human being. The robot will not be just another machine or a more sophisticated version of R2-D2 or C-3PO, in this view, but one who looks and feels like those we know as persons. Prostheses are becoming more and more realistic, and transplantation techniques are incorporating sophisticated technology that blends the artificial with the human. We now rely upon bioengineered body parts for breathing, cardiovascular, and renal functions, as well as substitute joints and prosthetics for extremities. Eyes with which to see and ears with which to hear, as well as skin-like material that senses hot and cold, is flexible and sensitive to touch, are already being developed. A cyborg may therefore look, think, feel, and talk much like its human creator.

Just how the downloading of consciousness will be achieved is a matter of considerable debate. That an equivalence of human intelligence might be imagined is one thing; getting it done is another. A more advanced stage of computer development will make possible more rapid and extensive computation, but it is less certain how computation correlates with consciousness. For all the experiments and developments over the past thirty years, no demonstration has yet been made that the two are identical.

Momenta toward the Future

The momentum toward a future dominated by cyborgs is driven by several forces: the medical mandate to seek the health and longevity of people; the nature of technology; and the fear of death, or the hope for immortality. The three forces obviously interact and enforce one another to create what is, for all practical purposes, a powerful and perhaps irresistible force.

The Medical Mandate. Beneficence is a grounding and guiding action principle for medicine. Physicians serve the well-being of patients and are committed to improving health and longevity. Impressive advances in the prevention and cure of disease and disability have taken place in the last half-century. Media portrayals of innovative medicine have contributed to unrealistic patient expectations and created a sense that medicine and health care are simply another commodity in a world of enormous options from which to choose. A consumer ethic creates an insatiable demand for new medical techniques and technologies. People seem to expect a quick fix for every malady. They insist on exotic procedures, products, or interventions whether or not they are of demonstrable medical benefit.[20]

Economic forces feed the frenzy since there is a great deal of prestige and wealth at stake for the healthcare industry. Insurance companies, pharmaceuticals, the bioengineering industry, healthcare providers, and a host of related services stand to gain financially from an expansion of new types of medical care. Skyrocketing costs result from fierce competition, expensive research and development, and the copyright or patent protection of trade secrets and products.

The net result is an atmosphere in which the continued interest in and development of bionic parts will undoubtedly continue and even increase exponentially as successful interventions are reported.

The Technological Imperative. Technology is not guided by moral or ethical mandates, as such, but provides market products in response to consumer demands or engineering possibilities. Advances in medicine are tied to new discoveries and innovative interventions made possible by new technologies. The problem is that technology tends to take on a life of its own, becoming autonomous and lending itself to little control by what are generously referred to as its human masters.

Further, technology tends to create its own ethic, the first rule of which is that if it can be done, it ought to be done.[21] Rodney Brooks is rather sanguine about the process of moving toward the development of cyborgs. He says it is simply a consequence of standard practices among scientists who move from analysis to synthesis, from theoretical hypothesis to technology and biomedicine. Cyborgs, he says, are simply a matter of bringing about new realities as a result of following standard scientific procedures. Wedding silicon and steel to biological matter is already commonplace—a fact of life in the world of the medical-industrial complex.[22]

In short, the questions for Brooks and other scientists are a matter of technology, not ethics. An ethos of technological imperatives is not inclined toward raising questions of metaphysics or values. What is right is what works; what is wrong is what hinders progress or injures people in the

---

[20] See Callahan, "Rationing Medical Progress," 1811.

[21] Jacques Ellul, *The Technological Society*, trans. John Wilkinson (Philadelphia: Pilgrim Press, 1969) 135ff. See also his *To Will and to Do*, trans. C. Edward Hopkin (Berea, OH: Pilgrim Press, 1969) esp. chapter 11, "Technological Morality."

[22] Brooks, *Flesh*, 185.

process of attempts at medical therapy. Such questions operate on the level of what is functionally beneficial or medically helpful.

Fear of Death, Hope for Immortality. Few dynamics are stronger motivators of human action than the fear of death.[23] The drive for new technologies is no exception. The artificial heart has a strong appeal as one way to cheat death of its prey.

Cyborgs thus provide a hoped-for path to immortality.[24] Kevin Warwick believes the cybernetic age is the next stage of human evolution. "The potential for humans if we stick to our present physical form," he says, "is pretty limited," but cybernetics offers a more promising and personal option. He welcomes further developments and says he can't wait to become a cyborg, primarily because it will enable him to live on without a mortal body.[25] Ray Kurzweil also predicts a future in which humans gain near immortality by blending robotic technology with the central features of humanity.[26]

Playing God and Apocalyptic Outcomes

Not everyone is happy about the biotechnical prospects for the future, however. Critics see the new changes as important not only for the novel entities they make possible, but for their impact upon the very being of humanity. The question is whether we know what we are doing and have adequate intellectual and spiritual foundations, that is, the moral wisdom, to do so responsibly.

The monumental shift toward a world of cyborgs creates a fundamental disquiet about our being human. People are willful and prideful. They are also self-loathing. Incapable of accepting the givens of being human, they seek to modify and alter the very being they profess to love as the apex of creaturely existence. Thoughtful objections should be considered before causing fundamental alterations to this creaturely embodiment of millions of years of evolutionary history.

Strong objections to the development of cyborgs can be found in the writings of Mary Midgley, Francis Fukuyama, and the new Luddites. They raise insightful questions, which represent those who believe technology is taking people in the wrong direction. Their perspectives vary widely, and their concerns reflect different philosophical approaches. They are united,

---

[23] See Ernst Becker, *The Denial of Death* (New York: The Free Press, 1970) 265ff.
[24] Brooks, *Flesh*, 204.
[25] See www.kevinwarwick.com and *Newsweek* (January 16, 2001).
[26] Ray Kurzweil, *The Age of Spiritual Machines* (New York: Viking Press, 1999).

however, in their concern for what it means to be human and the nature of the future being created. What follows is a summary of their central arguments with critical reflections on the points they attempt to make. The question is whether the objections they raise are adequate to hold back the momentum toward the future they find so objectionable. At a minimum, further discussion might be provoked around the particular points isolated for consideration.

Dame Mary Midgley poses philosophical and ethical questions about the expansive reliance upon technology and its impact on our humanity. Two realities underlie the human intellectual and spiritual venture with scientific design, she says. One is the alienation of the engineer from the entity designed. The engineer stands over against or outside the car that does not share his nature.[27] In short, the technological process expresses a Cartesian dualism, which separates the actor from what is acted upon; it is alienation at the most basic level of existence.

The second operating assumption of technology, she says, is that there is no transcendent norm by which to guide and judge ethical decisions. Postmodern thought has dropped the notion that anyone or anything is in charge of the evolutionary process. Randomness and chance are now the operative model. Science has been radically secularized but now lives by perpetuating the myth of technological wizardry and salvation by technique.

Further, the ambitious plans outlined for the future of bioengineering require a type of omniscience and omnipotence of which people are incapable. Only God was once thought able to control the vagaries of nature and history. Now people propose that they are the arbiters and designers of human destiny. God has now been dismissed as a factor in the human future.

The ethical downside to following the god of technological beneficence is also considerable. Enormous *hubris* is in evidence when people claim that they may now design their own future, Midgley says. Sounding a similar note, George Annas charges that the great evil of cloning, for instance, is that of depriving one of the "existential right to...ignorance of facts about his or her origin that are likely to be paralyzing for the spontaneity of becoming himself or herself."[28]

Questions of justice also surface in a new and more sinister way, according to Midgley, since those making decisions and wielding power are

---

[27] Mary Midgley, "Biotechnology and Monstrosity," *HCR* 30/5 (September/October 2000): 12.

[28] G. J. Annas, "Why We Should Ban Human Cloning," *NEJM* 339/2 (July 9, 1998): 124, citing the philosopher Hans Jonas.

certainly not just anyone or everyone. The technologists, the engineers, the "elite" will be the designer and those who control outcomes as to social divisions between the advantaged and the least advantaged. The technological future is not a democracy, but an oligarchy. Jeremy Rifkin thus charges that the scientist-engineer is "playing God" as the human is blended with the technological.[29]

Francis Fukuyama opposes the development of cyborgs and the new biotechnology on distinctively religious grounds. He pits natural law theology against secular science and predictably finds them incompatible. Cyborgs are a threat to humanity itself, he says, and portend "a post-human future." He fears the power of people who claim to be free to shape their own behavior and to engage in self-modification since doing so violates what he calls "human nature."[30] He fears that genetic engineering could undermine the "natural order" and the "universal essence" of what it means to be human. We have an obligation, he says, to preserve that "stable essence common to all human beings."[31]

Fukuyama believes there is an irreducible minimum that defines what it means to be human, what he calls "Factor X," which means it cannot be precisely defined. Even so, he says, the essence of the human cannot be reduced to moral choice, reason, language, sociability, sentience, emotions, consciousness, or any other single quality.[32] Such qualities are also inadequate as a ground for defining and defending human dignity, he says. Further, even if they could be downloaded or otherwise introduced into robots, such features would not make them human. Human beings are a great deal "more than a complicated machine that can be made out of silicon and transistors as easily as carbon and neurons," he says. People should thus protect the full range of their complex, evolved natures against attempts at self-modification by future advances in biotechnology.[33]

Fukuyama is also critical of the medical mandate to relieve pain and suffering as if it should trump all other human purposes and objectives. Nor should the moral obligation to heal be used to justify genetic modifications as a way to bring about human wholeness, he says. He even opposes psychotropic drugs, believing they alter the effects of genetic inheritance.

---

[29] See Jeremy Rifkin, *The Biotech Century* (London: Gollancz, 1998) 197–98.

[30] F. Fukuyama, *Our Post-human Future: Consequences of the Biotechnology Revolution* (New York: Farrar, Straus and Giroux, 2002), 128.

[31] Ibid., 159.

[32] Ibid., 171.

[33] Ibid., 172.

Because they are "a harbinger of things to come," they should be resisted. Ironically, he finds the mentally ill and embryos to possess "the full range of our complex, evolved natures," while those who are homosexual do not.[34]

Fukuyama argues for an "ontological status of emotions," as given in individual human beings, and that an "ontological leap" occurred somewhere in the evolutionary process to explain the absolute value of human beings over other creatures. Human dominion would certainly extend to mastery over cyborgs, he argues. He says that the special status enjoyed by people is because of "the human soul (which) is something directly created by God." For him, "theories of evolution which...consider the mind as emerging from the forces of living nature, or as a mere epiphenomenon of this matter" are "incompatible with the truth about man." He thinks scientists are unable to provide grounds for the dignity of the person or explain the mystery of the soul. Somewhere in evolutionary history, he says, "a human soul was inserted into us in a way that remains mysterious. Science cannot discover what the soul is or how it came to be."[35]

Fukuyama's opposition to cyborgs is thus based on his beliefs about certain religious teachings, which he finds irreconcilable with science and its claims to truth based in data drawn from nature or experimentation. Insofar as technology or genetic manipulations attempt to modify the givens of nature, he argues, they are to be resisted.

The New Luddites have very different reasons for opposing the development of cyborgs than the philosophical or religious objections of such people as Midgley and Fukuyama. This influential movement is comprised of scientists and others who have been deeply involved with developing the new technologies. They were once something like technopians, but have become critics who fear the aims and designs that are being imposed on the future. Ironically, they believe sophisticated robotic life forms are going to be developed and accept some responsibility for contributing to those possibilities. Such developments are hardly to be celebrated or supported, say the new Luddites, since they constitute a considerable threat to human well-being.

The problem, as they see it, is that the ethical or moral sensibilities of the future will also shift dramatically. Cyborgs may one day question their human creator and cease to be governed by the ethical restraints imposed upon them. They may want to dispense with their "gods" (their human

---

[34] Ibid.

[35] Ibid., 161.

makers) and get on with the business of creating ever new and more amazing creatures—more like themselves.

Bill Joy, inventor of Sun Microsystems and co-designer of three microprocessor architectures, SPARC, PicoJava, MAJC, and several implementations, calls this "the new Luddite challenge."[36] These engineers and scientists have been on the inner circles of technical know-how, but are now reacting to the utopian visions they see driving us toward an undesirable and terribly threatening future.

Visions of a future in which humans exploit technology for their own advantage have a dystopian possibility, of course. People may wind up at the mercy of the machine. The masses are already superfluous, a useless burden on the system dominated by technique and technology, Joy points out. Only a relatively small number of people are capable of maintaining, using, or dominating sophisticated devices and computers that now drive most every aspect of modern civilization. Given enough time and further developments, however, even the elite, who now retain some control over technology, will be at risk. The problem is that the machines will become so complex that human beings will be unable to make them intelligently. Only computers with humanlike capacities of imagination and intelligence will be able to make better computers. At that stage, cyborgs will be effectively in control.[37]

The new being to emerge will be a product of the combined powers of genetics, nanotechnology, and robotics (GNR), according to Joy and others of the new Luddites. The momentum toward the future has taken a quantum leap that makes the future of such a vision not only possible but much nearer than we had imagined. Joy believes that by 2030, machines will be built in quantity that are a million times more powerful than the personal computer of today. The prospect for completely redesigning the world is upon us. The replicating and evolving processes that were once confined to the natural world are now in the realms of human endeavor.[38]

The prospect of cybernetic creatures is precisely what distresses the new Luddites. Their fear is that cyborgs may replace the human species or that they might come close to achieving near immortality by downloading human consciousness. The power of self- replication through this futuristic

---

[36] Bill Joy, "Why the Future Does Not Need Us," http://www.wired.com/wired/archive/8.04/joy.html.

[37] Kurzweil, *Spiritual Machines*, cited by Joy, "Why the Future," 2.

[38] Joy, "Why the Future," 6.

blending of genetics, nanotechnology, and robotics is, for him, a clear threat requiring a radical but commonsense response.[39]

For such reasons, Joy argues, we cannot simply do science and not worry about the ethical issues at stake in the emerging world of the future. Joy is not a Luddite in the Theodore Kazinski[40] mold proposing that the most radical of measures be used to stop technology in its tracks, but he can imagine the day will come when he would be morally obligated to stop work that leads inevitably toward a new world of cyborgs.

Guidelines for an Ethical Response

Ethical issues in the new world of cyborgs and cybernetics are numerous and multifaceted. They can be found both in the interface of the physician-patient relationship and on the horizon of the future being developed in the laboratories of biotechnology. Bill Joy and other new Luddites stand as reminders that people are not always in control of their creations and that complicated beings challenge the limits of human wisdom. Critical reflections on the cyborgian prospects for the future and of the critics briefly examined above raise certain points that merit emphases as guidelines for the humanities as the biotech possibilities are critically engaged.

First, a Christian or humanitarian ethic requires an active engagement with the issues on behalf of those entrusted to our care. Passivity in the face of rapid and confusing change is ethically irresponsible. A thoughtful reflection on the nature of what it means to be human and what being human requires or prohibits in relation to technology are all-important questions for the new realities of the future. The future is open, and those concerned about the future will critically engage the destructive trends of the present for the sake of a future worthy of human dignity.[41] A moral resolve not to trust the managers is easier made than implemented, of course. People of science should be schooled in the humanities in order to be sensitive to the temptation to manipulate and coerce those who are subjected to their management skills and technological prowess. The first

---

[39] Ibid., 9.

[40] Kazinski is the scientist who killed three leading technologists by mailing each a letter bomb. He was finally arrested when his brother recognized certain traits in materials he had written. Kazinski's essay (available online) contains strong arguments against computer technology. It has become something of a bible for a cult following.

[41] Jürgen Moltmann, *Science and Wisdom*, trans. Margaret Kohl (Minneapolis: Fortress Press, 2003) 6.

line of control over technology must be through the actions of morally sensitive leaders in medicine, science, and industry.

Second, a critical but integrative approach is needed between science and religion. Neither all-out opposition nor unthinking isolation of either from the other will enable a comprehensive vision of and response to the larger issues toward which science is moving the world. Midgley's concern about "playing God" is a point widely made among religious people who seem to assume they are protecting territory belonging only to God. The point is well taken but problematic as a norm for governing science and technology.[42]

For one thing, the religious meaning of playing God is related to moral and religious judgmentalism, not doing creative things with technology. Science will and must "play God," if by that we mean engage and attempt to defeat the hindrances to human development and health. The ethical task is to work with God toward a desirable future.[43] Being more specific and insightful than using territorial generalities about who is and is not "playing God" will be required if ethics is to be taken seriously.

Third, the temptation to allow fear to become the basis for a humanitarian criticism of science needs to be recognized and resisted. Most thinking people are more than a bit nervous about what the future may bring, but fear robs the mind of creativity and deprives the moral actor of the imaginative possibilities that are available for positive response. More is needed than fear-based opposition in the face of the challenges of technology wed to humanity. It is true that the future may find us woefully unprepared to confront the consequences of our own *hubris* and will-to-power in misguided applications of biotechnology. The world of the future may be widely populated with cyborgs. Some will welcome that possibility in their pursuit of immortality, or the future may be a struggle for survival against those created in the image of the human but who have become the archenemy of people. Neither of those prospects should be sufficient for critics to launch an all-out war to ban such developments.

A starting point may be to acknowledge that *hubris* and will-to-power are found in technologists as well as their critics. No one is exempt from this common human malady. Further, hysterical social movements are created

---

[42] See, for instance, T. Peters, *Playing God?* 2d ed. (New York: Routledge, 2003) 11–14, for a helpful description of various meanings of the phrase and an insightful discussion of the stakes in the debate.

[43] Paul D. Simmons, *Birth and Death: Bioethical Decision-making* (Philadelphia: Westminster Press, 1983) 228.

by those who fashion scenarios of extreme threat based on misinformation and fears of developments that are unlikely to materialize. Only rarely will the most threatening scenarios become a reality. Science-fiction and horror films play on the imagination and its capacities for flights of fancy or a retreat into realms of fantasy. They have only a limited, sometimes surrealistic, relation to reality, however. Being scared out of our wits seems to be something people enjoy. Unfortunately, many people take the graphic portrayals of science fiction as realistic threats. A more objective and reasoned assessment of futuristic scenarios is necessary if ethics is to provide positive guidance for technological developments.

Fourth, a responsible ethic will give careful attention to the multidimensional meanings of "the human" that figure so prominently in the dialogue between science and religion. These include the fear of death, the meanings of consciousness, the limits of technology, and distinctively religious terms such as "soul." Critics are undoubtedly right in saying that technopians are driven as much by a fear of death and the frustrations of mortality as by a realistic assessment of the prospects of downloading human consciousness. Insofar as cyborgs are made in the image of the human, they will also have something like a fear of death. Being able to live longer is no cure for anxiety about the cessation of one's being at some undesirable or predictable point in time. The notion that cyborgs will displace or at least make war on people is another evidence of death anxiety. The truth is more likely that cyborgs will be sophisticated and numerous, but they will remain limited as to the scope of their abilities to dominate people. Further, the prospects for enabling a robot to achieve transcendent self-awareness are extremely limited. In this arena, the imagination is more active than the technology is capable.

Fukuyama is right to say that there is something "more" to being human than can be explained by a clever organization of silicon and transistors, carbon and neurons,[44] but his use of "soul" as an explanation for the difference in being human is more than a bit problematic. It seems more a pious dodge than a meaningful claim when he says that sometime in the distant past a mysterious action inserted a soul into persons.[45] That "science cannot prove" a religious claim about the soul is true, but neither can

---

[44] Fukuyama, *Our Post-human Future*, 168.

[45] Ibid., 161. Fukuyama's notion of Factor X as a way of speaking of a human essence poses enormous problems. He wants to preserve some "universal essence" of what it means to be a human being. Incredibly, he excludes those who are homosexual while arguing that the mentally ill and embryos fully possess that essence (see p. 153).

Fukuyama prove the existence of a soul. The notion of "soul" is far too slippery and ill defined to use as a norm for science or technology. What we can use are the very features of human personality that Fukuyama rejects—namely, capacities for moral choice, language, sociability, sentience, emotions, reasons, and consciousness.[46] In that sense, cyborgs may very well be human; that is, they will be endowed by emotions, motives, and capacities that we identify as central to our own humanity.

The relation between science and religion needs a more careful analysis than what Fukuyama provides. Among other problems, his approach to public policy would violate the constitutional rule that "Congress shall make no law respecting an establishment of religion." He wants public policy to protect ecclesiastical doctrine, as if religious "truth" is not subject to the usual rules of critical analysis. Unfortunately, Fukuyama embraces an ancient tradition that is woefully inadequate to engage the critical questions of science and the future.

A fifth guideline is that there is no implicit threat to people in the fact that sophisticated cyborgs may be developed. Cyborgs will be an extension of our humanity, much as the automobile and other artifacts are extensions of human personality. People will likely develop an emotional attachment to cyborgs, much as we do to other possessions and pets.[47] Fukuyama's statement that he could destroy a Spock-like creature without any remorse or second thought[48] is surely misguided and irresponsible, not to mention ethically problematic. Such comments betray a lack of insight into what it means to be human in relation to other creatures, especially those made in our image.

Sixth, Christian ethics need to spell out the grounds for norms that are proposed for guiding science. There may be a "transcendent norm" that should govern the aspirations and actions of science, as Midgley says, but how we propose norms that have some prospect for acceptance among reasonable people still remains to be settled. Even if we grant the notion that God is the transcendent norm, how does that translate into moral action guides for the scientist? Who is to say that making a cyborg violates God's norms, and how would they establish that they know God's mind on this matter? History is replete with people who feared what science was

[46] Ibid., 171.

[47] What I have in mind is something like the relation of "Wilson" to the tragic hero in the movie *Cast Away*. A volleyball became the beloved companion of the pilot who managed to survive on an isolated island. "Wilson" was a volleyball, not a person. But "he" was given a name and took on a value far beyond the fact of his limited and even non-cognitive state.

[48] Fukuyama, *Our Post-human Future*, 169–70.

doing as displacing or angering God. Surgery was opposed, as was the very notion that earth was not the center of the universe, or people the central object of God's providence and care. We can affirm the limited wisdom with which we work without believing that we should turn to clerics for the norms by which science should be governed. Whatever wisdom there is, it is a shared wisdom. Science and religion have a great deal to offer to one another and to learn from one another. The notion that pious generalities like the fear of playing God or that we are simply acting out of *hubris* hardly provides the solid guidance necessary for the tough choices in technology and science.

Finally, critical thinking about and a strong commitment to issues of justice will be basic to a responsible ethic of and for the future. Critics of the new technologies rightly focus the question of justice in the use and applications of the powers of science and biotechnology. The new Luddites represent an emphasis on conscience in the technological sector in that they are concerned about the applications and consequences of the new technologies. Recognizing what Dame Midgley calls "issues of justice" is neither the result of paranoia nor an outgrowth of some outrageous conspiracy theory. Society is increasingly pyramidal, and the gap between rich and poor is expanding daily. People at the pinnacle of wealth and power make decisions and the laws that protect their own privilege at the expense of those who comprise the major portion of the world's population. Those making the decisions and wielding power are all too willing to believe they are and should be the elite governors without any moral obligation toward the masses. The technologists, the engineers, the political "elite" will be the designer and those who control the processes and benefits of science. Shared power and ways to govern the distribution and allocation of resources are continuing and vexing problems, requiring the committed energies and thoughtful responses of persons sensitive to humanitarian concerns.

Conclusions

We can reasonably predict that the twenty-first century will be a world in which the differences between the artificial and the natural will be increasingly difficult to discern. Machines will be made in the image of their human creators. They will be intelligent, thinking, calculating, feeling. And they may well be concerned with survival, replication, control of their environment, and dominance over competitors, just as people are.

The great irony in the uses of technology is that we are also changing our humanity. Science can never do just one thing. The connections between various possibilities are not always immediately apparent, but an intuitive moment will make the application, and a new event in time and space will be produced. People are already becoming more cyborg-like, more dependent upon an elaborate and complex health support system, more committed to their artifacts and technological extensions, and less able to live as self-contained beings.

Basic to this monumental shift is a fundamental disquiet about what we are doing to our humanity, to our being human. People are willful and prideful. They are also self-loathing. Incapable of accepting the givens of human existence, they seek to modify and alter the very being they profess to love as the apex of creaturely existence. The advent of cyborgs may well provoke a more careful and insightful appreciation of who we are and how we should pursue and preserve a worthy future for all creation.

Chapter 8

ETHICS AND CTA: MIGHT WE HAVE YOUR FACE?

Scientists and ethicists are virtually unanimous in saying that there are some things that should not be done to human beings. The late Paul Ramsey stated the principle strongly: "There are any number of things that we can do that ought not be done. There are some things that we cannot morally get to know."[1] There is more agreement about the moral principle than on its application to particular procedures, however. Ramsey was writing in opposition to in vitro fertilization (IVF). Sperm and egg were being brought together in a petri dish as a step toward assisting women to deal with infertility. He was convinced that the procedure should not be done, even if it could be, since it violated certain moral values that he believed were non-negotiable. The ethical and medical climate of opinion has shifted considerably away from Ramsey's staunch opposition to IVF. Even so, many of those who disagreed with him about the moral acceptability of IVF nevertheless accepted in principle the point he had articulated so well: there are some things that can be done that ought not be done.

Transplants involving face and hands raise that issue in a pointed and powerful way. Allotransplants, or vital organ transplants from one person to another, now have solid support and even moral praise in the medical and religious communities. Both cadaver and living donor transplants are widely relied upon to meet the needs of patients who are dying because of the failure of one or more vital organs, but what of transplants involving non-life-saving organs?

Composite Tissue Allotransplants (CTA)

Such questions have lately made international news. A team of French surgeons, led by Bernard Devauchelle and Jean-Michel Dubernard,

---

[1] Paul Ramsey, "Shall we 'Reproduce'? I. The medical ethics of in vitro fertilization," *JAMA* 220 (June 5, 1972): 1347.

performed a partial face transplant in November 2005, grabbing headlines around the world. Surgeons in Cleveland and Louisville have also announced they are searching for qualified candidates for face transplants,[2] so the face has now been added to the list of CTA that have been completed or are being planned for the near future. Already those who ask for people to sign their organ donor card have begun to ask about such body parts. Some have been asked, "May we have your face?" Getting used to the idea will take some time, but providing parts from the body for allotransplants will undoubtedly increase in the near future.

An allotransplant is an organ transplanted from one person to another. A xenograft is an organ that is transplanted between species, as a heart from a baboon to a human infant, as in the infamous Baby Fae story.[3] "Composite tissue" refers to the variety of bodily tissue involved, including bones, muscles, sinews, blood vessels, nerves, and tendons. A solid tissue transplant refers to organs like kidneys, livers, and hearts, which are composed of basically the same tissue throughout. A hand or face is composite tissue; a liver is solid tissue, but the distinction of critical importance is that the solid organ transplant is designed to give a person an extended life. One cannot live without a heart, kidney, liver, etc., but a composite tissue organ does not prolong a person's life, though it may be personally desirable. A hand or face transplant is a surgical accommodation to the loss of a significant bodily organ, but it is cosmetic, or a matter of correcting a disfigurement. And thus the issue of importance to ethics: transplanting any organ requires the use of anti-rejection drugs, which are life-threatening in themselves. Does the medical doctrine of "do no harm" prohibit a transplant that may cause a premature death without the compensatory benefit that one's life is at stake?

There are people of influence and medical expertise who oppose such transplants precisely because the threat to life is thought to constitute a burden that is too great to justify the procedure. This chapter examines the critical ethical issues at stake in the debate around several points: (1) the principles of medical ethics; (2) the procedures employed in procuring informed consent and protecting research subjects; and (3) the arguments set forth both in opposition to and in favor of such transplants. My interest is not simply to outline the issues but to suggest a resolution of the debate.

[2] O. P. Wiggins et al., "On the Ethics of Facial Transplantation Research," *American Journal of Bioethics* 4/3 (2004): 1–12.

[3] Baby Fae was born dying with hypoplastic left heart syndrome. Her destitute mother gave consent for Dr. Leonard Bailey to give her the heart of a seven-month-old baboon. The procedure was entirely experimental. Baby Fae died 21 days later.

My position is one of qualified support for proceeding with CTA programs. By my analysis, the burden of proof falls on those who would deny this procedure to patients who have suffered the loss of a major limb or body part.

Codes and Conundrums

The concerns regarding risks to a patient that are life-threatening cannot be lightly dismissed. One of the few types of research with human beings that is explicitly forbidden by the Nuremberg Code is that in which "there is *a priori* reason to believe that death or disabling injury will occur" (See Appendix IV, para. 5.). That norm seems clear and is widely accepted. Rarely, if ever, is a medical experiment conducted when it is known that it will likely result in death or injury to a patient. Such negative outcomes are few and far between.[4] Some terribly adverse events, including death, may be inevitable, but they are hardly foreseen or intended.[5] Those who drafted the Nuremberg Code had in mind such horrific Nazi experiments as those of Joseph Mengele with Jews in the North Sea,[6] or Hans Eppinger with Gypsies at Dachau.[7] The reasoning of the Nazi researchers was both dispassionate and distorted. Since the Jews and Gypsies were condemned to die anyway, why not try to learn something from experiments with them? The aim ostensibly was to contribute to the store of medical knowledge, but the research was driven by ideology, not beneficence or compassion for patients. Knowledge about human tolerance for hypothermia was vital for military as well as civilian personnel, and Eppinger's desire to make salt water potable was certainly understandable if not commendable, but his experiments were designed to use victimized groups in order to advance a political agenda. Whether or not any useful knowledge was gained remains a matter of serious debate.[8]

---

[4] Leon Eisenberg, "The social imperatives of medical research," *Science* 198 (December 16, 1977): 1105–10.

[5] Jesse Gelsinger, eighteen, died from an experiment with gene therapy at the University of Pennsylvania, and Ellen Roche died in an experimental procedure at Johns Hopkins University.

[6] The Nazis used Jews for experiments on hyperthermia in the North Sea. See Arthur L. Caplan, ed., *When Medicine Went Mad: Bioethics and the Holocaust* (Totawa, NJ: Humana Press, 1992).

[7] "Infamy Haunts a Top Award," *Time* (November 26, 1984) 89.

[8] See R. L. Berger. "Nazi Science: Comments on the validation of the Dachau human hypothermia experiments," in Caplan, *When Medicine Went Mad.*

The Medical Betrayal

The fact that the most noble of medical motives can become distorted by ideology has given rise to the emphasis on biomedical ethics. Nuremberg and the Nazis may seem remote, but America has its own tainted history with the infamous Tuskegee syphilis experiments with African-American men in Alabama.[9] Moral values need to guide the use of procedures and technologies in medicine so important human rights are not violated and vulnerable patients injured or killed for the sake of the good of society or the advancement of medical knowledge. Considerable attention has thus been given to the need for protecting research subjects in recent years. The basic message is that care must be taken to ensure that the goals of medicine are governed by human values and that patient needs outweigh the physician's ambition or desire for notoriety or prestige.

Allotransplants: The Need and the Controversy

CTA programs focus the issue of patient protection with new urgency. Controversy rightly and predictably attends the introduction of novel medical procedures on ethical as well as medical grounds. The desirability of a transplant for persons who have lost a limb from injury or disease has long been recognized. The painting attributed to the Master of Los Balbases, Burgos, Spain, dated circa 1495, depicts Sts. Cosmas and Damian, the patron saints of medicine, appearing in a dream to physicians who have removed a cancerous leg and are replacing it with the leg of a dead Moor. Their dream of an allotransplant in many ways anticipated miraculous advances in medical science. Medieval surgeons had little understanding of the body's immune system, but they did understand the human need for bodily wholeness after amputation or loss of limb. That dream was not new; it had existed since people could think about the desirability of restoring a limb lost to accident, injury, or surgery.

Recent developments in medicine may be making the pursuit of that goal more realistic. Since the first hand transplant in Lyon, France, in 1998, there have now been a total of eight transplanted hands. That number includes a double amputee in France and another in Austria. Other transplants include persons who received knees, thigh bones, and a larynx, which included additional tissue and sections of blood vessels.[10] As of early

---

[9] See Gregory E. Pence, *Classic Cases in Medical Ethics*, 4th ed. (Boston: McGraw-Hill, 2004) 270ff.

[10] M. Strome, "Larynx transplantation," Second International Symposium on

2004, of the six persons who received CTA, eight grafts are still surviving. The first American hand transplant recipient celebrated the seventh anniversary of his surgery in January 2006.

The Ethical Challenge

The justification for an allotransplant has been challenged by those who question whether the transplant should be done at all, given the present state of medical knowledge. The issue is not whether it is possible technically, but whether it is permissible ethically. The question is not just what can be done *for* a patient but what is being done *to* a patient.

Severe complications related to the patient's immune system may accompany an allotransplant. The very nature of the procedure makes extreme caution necessary. A hand would be a composite-tissue entity, thus involving skin, muscles, tendons, bones, cartilage, nerves, fat, and blood vessels, each with its own degree of immunogenicity, or resistance to foreign bodies. Immunosuppressant drugs must be taken for the rest of the recipient's life in order to prevent rejection. The drugs themselves are dangerous. The issue is thus posed as to whether a patient—even a willing and eager patient—might ethically be allowed to receive such a transplant. After all, one patient in 100 will die from using the immunosuppressant drugs themselves.

Further, there is a higher risk of malignancy, infections, and metabolic disorders among transplant recipients. A higher rate of skin cancer, lymphoma, cervical cancer, and Kaposi's sarcoma is found among patients whose immune system is suppressed by drugs when measured against the general population.[11] Cyclosporine, the drug used to suppress the immune system and thus enhance grafting, may actually trigger the growth of cancer cells.[12] There is also a risk for developing post-transplant lymphoproliferative disorders (PTLD), which increases with the presence of a number of variables, including previous exposure to Epstein-Barr virus (EBV), and the use of antilymphocyte antibodies.[13] Other risks include bacterial, viral,

---

Composite Tissue Allotransplantation, Jewish Hospital, Louisville, KY, May 18, 2000.

[11] S. A. Gruber & J. A. Matas. "Etiology and pathogenesis of tumors occurring after organ transplantation," *Transplantation Science* 4 (1994): 87–104.

[12] "Transplant drug increases cancer risk," *Science News* 155/12 (March 20, 1999): 187.

[13] N. Basgoz, J. K. Preiksaitis. "Post-transplant lymphoproliferative disorder," *Infectious Disease Clinics of North America* 9 (1995): 901–21.

fungal, and opportunistic infections, the latter being more common later in the post-transplant period.

Such problems attend any transplant, of course, as they do major surgery of any type. Surgery is the art of doing very risky things to a patient with the hope of correcting defects and thus improving health or extending the life of the patient. The patient who receives a vital organ must also take immunosuppressant drugs for the remainder of one's life. The major ethical difference between a kidney or heart transplant and a hand transplant is that kidneys and hearts are life-saving; a hand transplant has to do with quality of life, but it is not life-saving as such. At least part of the resolution of the ethical challenge involves a type of utilitarian calculus: it is a matter of weighing risks against benefits.

Calculating Benefits vs. Risks

A position paper adopted by the Council of the American Society for Surgery of the Hand argues that the risks involved are so considerable that the procedure should simply not be done. For them, the ethical rule of *primum non nocere* (do no harm) is violated since the procedure has not yet proven sufficiently beneficial to justify continued exposure of patients.[14] Further, the International Federation of Societies for Surgery of the Hand (IFSSH) has recommended a cautious approach until more information on current patients is gathered and processed.

Other critics can also be found. Neil Jones said bluntly, "We do not think the risk-benefit ratio is in favor of the patient at the moment."[15] Among other things, he emphasizes the fact that, even if the organ is not rejected, there is a long-term prospect for developing cancer. He thinks suspending further CTA seems necessary in order to evaluate a patient's long-term prospects. There is a danger, he says, in emphasizing short-term gains that may cause a premature stamp of approval that ignores the long-term threats. German surgeon Gunther Hofmann, who transplanted knee and thigh bones for accident victims facing leg amputation since the mid–1990s, now supports a moratorium on CTA. He reported that, even

[14] W. P. Cooney and V. R Hentz, "Hand Transplantation—*Primum Non Nocere*," Position Statement of the Council of the American Society for Surgery of the Hand, *The Journal of Hand Surgery* 27A/1 (January 2002).

[15] Neil Jones, "Histological, immunological and functional comparison of nerve allografts immunosuppressed with cyclosporine, FK–506, RS–61443 and rapamycin," Second International Symposium on Composite Tissue Allotranspantation, Jewish Hospital, Louisville, KY, May 19, 2000.

though the femur transplants seemed at first to have succeeded, two of five transplanted knees failed because of rejection and were replaced by artificial joints. He expressed a lack of confidence in the transplant surgeons' ability to recognize and treat early signs of rejection and thus believes "it is necessary to return to the laboratory."[16]

Benefits: Patient Perspectives

The benefits of the CTA are harder to quantify since they are primarily related to quality-of-life considerations. Greater life expectancy can certainly not be assured since the patient's life will likely be shortened. Values are difficult to calculate in terms of percentages or odds when one turns to the patient to inquire about the positive health effects associated with having received a composite tissue allotransplant. Patient responses understandably tend to be highly, but not entirely, subjective. They speak from experience and deal with weighty matters of importance to them, such as relations with loved ones, intensely personal losses, and a sense of what it is worth it when an opportunity for restoration presents itself.

Timothy Heidler's larynx was destroyed when he rode his motorcycle under a cable strung across the road. When presented with the opportunity, he jumped at the chance to have a larynx transplant. He was incredulous about those who thought he should have been perfectly happy with the prosthetic electronic voice box that enabled him to communicate by using vibrations from his throat. His response to those who fear the drugs will kill him and that he should not have accepted the donated larynx is that he would do it again in a flash. The benefits of his new larynx can hardly be measured or understood by those who take their speaking capacity for granted. His new voice is "his," he says. And others in the family tell him he sounds like his dad. The personal and psychological payoff for being rid of a non-human-sounding box and regaining a human voice are simply immeasurable, he said. Asking him to settle for the prosthesis is simply not what he is all about. He is determined to fight back, to attempt a remedy to the tragedy experienced.[17]

---

[16] G. O. Hofmann, "Knee and femur transplantation," Second International Symposium on Composite Tissue Allotransplantation, Jewish Hospital, Louisville, KY, May 18, 2000.

[17] Timothy Heidler, "Patient perspectives," Second International Symposium on Composite Tissue Allotransplantation, Jewish Hospital, Louisville, KY, May 18, 2000.

Matthew Scott, a New Jersey EMS instructor who received a hand transplant in Louisville, had never been satisfied with the prosthetic hand he had received after a firecracker (M–80) accident. His "yearning to be normal" and his sense of loss regarding touch toward his children and wife created a feeling of deprivation that the hand transplant helped correct. For him, being able to hold his children and touch his wife make the risks "worth the gamble." He says he is "far better off than with the prosthesis." He, too, would do it all over again.[18]

Calculating risks versus benefits is seldom simply a matter of stating the risk factors in medical, that is, statistical, terms. Knowing the risks or threats is necessary but not sufficient in the calculus of informed consent. This caveat is important both for the patient and as responsible surgeons struggle with the question of when to move forward with this innovative procedure. Patient perspectives are vital as risks are weighed against possible benefits. Both Heidler and Scott expressed strong feelings about the prospects of being denied a procedure that they believed would be not only acceptable, but desirable to them. Against the ethics of doing no harm is the ethical problem of withholding what may be perceived as a beneficial and therapeutic intervention. Beneficence is a stronger moral obligation than that of doing no harm. Beneficence takes the form of an active effort to correct an injury or harm that has enormous importance for one's mental and personal sense of well-being. Doing no harm is certainly morally obligatory, but it should not become an excuse for doing nothing when acceptable procedures are possible and of proven benefit.

Ethics, Processes, and Procedures

Even so, caution is clearly in order since the risks associated with CTA are so high. Professional ethics mandates that care be exercised in order to avoid a cavalier attitude toward innovative procedures that may cause more harm than good. The Louisville hand transplant team thus engaged in a rather complex process of evaluating whether and when to proceed. The first step in the process decided on four ethical issues to be addressed: (1) ensuring public awareness of the prospect for moving ahead with an innovative surgery, (2) evaluating risks and benefits to the recipient, (3) selecting appropriate recipient and donor, and (4) informed consent.[19] In addition,

---

[18] Matthew Scott, "Patient perspectives," Second International Symposium on Composite Tissue Allotransplantation, Jewish Hospital, Louisville, KY, May 18, 2000.

[19] W. C. Breidenbach et al., "The ethics of the first human hand transplantation,"

two overarching issues dealt with the scientific research on allografts and the expertise, or "field strength," of the transplant team.

Consultation and guidance during the initial stages was provided by Dr. Mark Seigler of the MacLean Center for Clinical Medical Ethics at the University of Chicago. While not a member of the surgical team, he was named "ethicist-observer." His guidance as to the criteria to be assessed and the procedure employed closely followed protocols governing a living donor liver program at Chicago that had been developed by Dr. Francis Moore.[20] Seigler also wrote a paper on the ethics of the hand transplant proposal,[21] which was widely disseminated and discussed.

The transplant team also reflected upon professional standards of medical ethics as contained in the Declaration of Helsinki, the Nuremberg Code, and the standards established by the Council on Ethical and Judicial Affairs of the American Medical Association.[22] The team placed high priority on professional and public discussions. Symposia on CTA were conducted in Louisville, Vancouver, and Chicago that brought together internationally recognized experts in hand surgery, transplantation, immunology, and bioethics. Comment and dialogue were invited on every aspect of the procedure, even before a protocol was submitted to the Institutional Review Board (IRB).[23] The hope was to generate discussion and gather information that might be beneficial by showing a need to modify or correct procedures or surgical approaches.

My first involvement with the hand transplant project occurred when Dr. Warren Briedenbach, a member of the team of surgeons for the Matthew Scott transplant, spoke to the bioethics committee of the Jefferson County Medical Association[24] to discuss the project and deal with questions that might be posed. Dr. Briedenbach's paper had been distributed to members of the bioethics committee on which I happen to serve. I had no relationship to the hospital or the transplant team and did not know

---

1998, unpublished, 3.

[20] Francis D. Moore, "Three ethical revolutions: ancient assumptions remodeled under pressure of transplantation," *Transplantation Proceedings* 20 (1988): 1061–67. See also, F. D. Moore, *JAMA* 261 (1989): 1483.

[21] Mark Seigler, "Ethical issues in innovative surgery: should we attempt a cadaveric hand transplantation in a human subject?" *Transplantation Proceedings* 30 (1998): 2779–82.

[22] Council on Ethical and Judicial Affairs, *Code of Medical Ethics: Current Opinions with Annotations 2006–2007*, American Medical Association, Chicago.

[23] Briedenbach, "Ethics," 3.

[24] Now called the Greater Louisville Medical Society.

personally any member of the team prior to that time, but I became deeply involved in the discussion since I had prepared written comments raising questions about certain points in the paper. Some time later, another member of the transplant team, Dr. Gordon Tobin, chair of the plastic surgery department, and I shared a television interview dealing with the hand transplant. I also led a ground rounds lecture with Dr. Briedenbach on the ethics of the hand transplant.[25]

High marks should be given the surgical team for the patience and dedication with which it went about seeking ethical input. Every effort was made to make whatever corrections seemed warranted before requesting approval to proceed from the IRB. A deliberate effort was made to be open to criticism and alternative points of view that might persuade them not to proceed with the hand transplant.

Experimental medicine pushes at the frontiers of traditional ethical formulations and established medical procedures. The research physician works in that ethical penumbra created by a commitment to developing more effective therapeutic interventions. The researcher has a moral obligation to seek new cures, new devices, and new and more effective therapies—all of which are impossible without experimentation and/or introducing innovative procedures. Along with the possibility of developing new therapies goes the potential of unintended injury or death to the research subject. Pushing the boundaries runs the risk of injuring patients by moving too quickly with unproven therapies. Even so, the presence of risk does not rule out the research. The Declaration of Helsinki, which provides guidance for experimental procedures with human subjects, recognizes that biomedical research involves hazards.[26]

Ethical Concerns and Innovative Surgery

Traditional moralists tend to err on the side of caution, thus delaying or preventing more effective therapies from being developed or made available to the patient. The researcher, on the other hand, is tempted to err on the side of innovation, frustrated by the constraints of bureaucracy, regulatory requirements, rigid moral formulations, or religious objections.

---

[25] Warren Breidenbach, "Medicine at the Edges: Ethics and Innovation," Grand Rounds, 16th Floor Conference Center, Rudd Building, Jewish Hospital, Louisville, KY, August 20, 1999, Unpublished.

[26] *Declaration of Helsinki*. World Medical Association, 48th General Assembly, West Somerset, Republic of South Africa, 1996.

Two schools of thought tend to dominate the scientific discussion. The first might be called "launch when ready." Readiness refers to the attainment of technical expertise and that the patient's informed consent has been acquired. The second approach insists surgeons should launch only after widespread discussion, a consensus has developed among peers, and support has been given by the professional association. The basic question at issue is how much and what kind of consensus among physicians and/or the public should emerge before the surgical team should proceed. The first approach trusts the integrity of the physician and his or her intuitive guidance as to the proper moment to respond to human need with unproven medical techniques or technologies. The second approach tends to emphasize community responsibilities for protecting subjects and thus slows the process of moving from research to application.

British physician-researchers tend to favor the latter. Dr. Robert Edwards, co-founder of the in vitro fertilization program at Bourne Hall, England, was once asked whether he would engage in an innovative procedure without permission from the Warnock Committee, the National Committee on Bioethics in Medicine in Great Britain. The issue at the time was a procedure for blastomere separation for genetic analysis. Dr. Mark Hughes, a Houston geneticist, had worked with a process involving the removal of a single blastomere from a pre-embryo created in a petri dish. The blastomere could then be examined for a particular lethal genetic burden. Lesch Nyhan[27] had been manifested in a series of pregnancies by a couple presented to Hammersmith Hospital, London. Needless to say, they were terribly frustrated and increasingly desperate for a healthy child. The genetic tests resulted in distinguishing three groups of pre-embryos: the normal, the carrier, and those affected (by Lesch Nyhan). Only the normal were returned to the woman. A pregnancy was established, and a healthy boy was born to the couple. By the time of our conversation, such testing was also being done at Bourne Hall, but the question posed was whether Edwards would have engaged in such testing without the consensus represented by committee approval? He said proceeding without approval

[27] Lesch Nyhan Syndrome (LNS) is a rare but lethal genetic disorder among boys (never girls) involving malfunctions of uric acid that result in severe physical and neurological problems. The affected will be mentally impaired and lack muscle control. Disorders include extreme biting of fingers and lips, severe kidney problems, and stiffness in the joints. Central nervous system (CNS) difficulties result in involuntary writhing, facial grimacing, and repetitive movement of the arms and legs. There are treatments to alleviate symptoms of LNS, but there is no cure for the disorder. Life expectancy is 5–25 years.

was simply unthinkable. Even if he had been the first to perfect the procedure and even if in his own mind it was harmless to those involved and helpful to the patient, he would not have proceeded.

Going Public: How Much Debate Is Enough?

The Louisville hand transplant team adopted the slower path so as to deal carefully with all ethical questions that were raised. Was a time-consuming, national, and international discussion on hand transplants ethically required? Surgeons followed the suggestion of its ethicist-observer that a widespread discussion precede the CTA. The suggestion had the appeal of allowing all arguments pro and con to be heard and all objections considered over time sufficient to allow new insights or convictions to emerge.

Two substantive arguments against the proposal for broad public debate could be made. The first was that discussions about ethical issues are seldom conclusive, but they can be terribly time-consuming. As it turned out in the hand transplant case, the delay served primarily to exacerbate the debate and delay a procedure for which many people were waiting with keen anticipation. There are hundreds of people eager for a hand transplant, judging by the numbers who applied to the Louisville team. Far from settling the ethical questions, the two sides simply refined their arguments and hardened their positions. No new data or additional insights were sufficient to alter the assumptions and arguments being made. Definitive data would not be available until the transplants were actually performed and could be studied over time.

What was demonstrated is that ethical debate is a mode of raising questions and outlining arguments that seem to justify or forbid a particular pattern of thought or behavior. Opinions strongly offered tend to generate counter opinions. The dialogue typically generates a more refined position, seldom a change of mind. Meanwhile, all comers are invited to the debate. The anti-vivisectionists, the anti-technologists, the Stoics, the suffering-as-divine-punishment people, the fear-mongers, the jealous, and the ill informed all get their voice on the daylong or primetime talk shows. In the end, nothing was settled except the fact that people see the issues in very different ways, each of which has its own mode of reasoning and constituency of support.

The second argument against delay was that there was and is sufficient expertise, both technical and ethical, locally or regionally, to evaluate the various issues pertaining to the allograft. Whatever might have been controversial about the procedure was surfaced among those in the medical, philosophical, and religious communities in the Louisville area very early in the discussion. The need to extend the debate over a prolonged period of time was hardly self-evident, though it certainly had pragmatic and professional value.

At the pragmatic level, one benefit was that no one in other centers of medical research or ethics could complain that they had had no opportunity to comment. Professionally, the leaders of the hand transplant team are part of an international network of surgeons. As such, there are obligations both to the guild and the academy to share information and be open both to correction and encouragement. Innovative procedures introduced under the cloak of secrecy generate suspicion, if not opposition, perhaps for all the wrong reasons.

As it turned out, the effort for "public display"[28] had the effect of notifying the world of the intention to proceed with the allograft and thus encouraging others to beat them to the draw. Thus, the first hand transplant took place not in Louisville, but Lyon, France, where a double transplant has also now been performed. There are good reasons why the Louisville team *should have been* the first to transplant. The Kleinert and Kutz group enjoys an internationally recognized preeminence and is regarded as the standard bearer in hand surgery. Perhaps the rest of the world should have waited for them to make the move and in so doing announce to everyone that the time had come to move forward.

The expertise of the French team need not be disparaged when saying that the Louisville team should have performed the lead CTA. In a sense, the French effort was a commentary on frustration with the delay, or the motives may have been less honorable: perhaps they were determined to be first so that they will be mentioned when the history is written and reduced to a phrase or sentence.

They may also have shown more zeal to pursue an important medical goal. There is a point at which further delay simply increases frustration and extends patient suffering. The Louisville team may be judged by history to have been overly cautious, too deferential to critics, or too concerned about public opinion. Perhaps they were too inattentive to competitors who were less concerned about the niceties of professional ethics. There are patients to be cared for and procedures to be perfected. After the arguments have been made and considered and no objections have surfaced that provide a persuasive argument against the procedure, and no public outcry has been generated, the time has come to proceed.

---

[28] Seigler. "Ethical Issues...." *Transplantation Proceedings* 30 (1998): 2780.

Another negative effect of the lengthy public discussion was to make the hospital open to the charge that the primary motive was institutional self-aggrandizement. Dr. Moore, in guiding the liver transplant protocol at Chicago, had argued that "the ethical climate" might discredit and disqualify an institution if "capital gain, investor profit, institutional representation, surgeon ego, municipal pride, and chauvinism" displace the aim of contributing to those with personal medical needs.[29] That accusation surfaced in discussions about the prospect for the hand transplant as part of a more general concern with advertisements by prominent hospitals in the Louisville area.

The competitive economic climate in which medicine is now practiced constitutes a major ethical concern. Thus, the motives of those raising the question may have reflected a genuine ethical concern, but the charges against the transplant team seemed both misplaced and unwarranted. Public attention, that is, media coverage, was necessary to the goals of transparency, openness, and public involvement. To insist that no attention or credit be given the hospital would be to undermine the very reason for going public. That there may be some economic advantage to the sponsoring institution simply goes with the territory. It just so happens that the hospital at issue is also one of two major transplant centers in Kentucky.

Patient Preferences and Informed Consent

Ethically, the concern for the health and well-being of the patient must be the top priority for the transplant team. Every step in the process should be designed to assure that the benefits of the procedure are not exaggerated and that the risks are carefully explained to the recipient. There are three requirements that must be met for therapeutic experimental surgery to be ethical. First, background research studies must be sufficient to establish reasonable expectations that the procedure will benefit the patient. Second, there should be no alternative procedures available of known or equal benefit. Third, the patient must give consent that is free and informed.

The first requirement assures that the risks are minimized and solid prospects for patient benefit can be reasonably assured. The second recognizes that while prosthetic devices enable amputees to function in often satisfactory ways, they cannot serve to compensate for the sense of loss that is so highly personal and psychologically devastating. The third requires

---

[29] Siegler, Ibid.

careful attention to details in the process of securing the consent of the patient.

The known risks associated with CTA make it ethically imperative to provide all information that would prove "of material benefit to the patient in reaching a free and informed decision."[30] The patient is a partner in the decision to cross the line with an innovative procedure. Informed consent on the part of the patient is thus as critical to the ethics of the procedure as are the technical questions pertaining to physician competence and risk factors. The "covenant" between physician and patient is that a partnership of mutual trust, and respect will be fashioned in order better to assure optimal outcomes.

The very notion of informed consent is problematic, however, when a patient has been through a traumatic loss of a limb and researchers want to engage in a noteworthy procedure. The first problem is that treatment options are extremely complex and the patient is a layperson. How can a scientist explain complex issues in immunology and tissue rejection to one who is not trained in medicine? Even scientists in other fields of medicine have difficulty communicating across lines of specialty and various medical disciplines.

Second, the patient is eager to hear the "promise" of benefits and perhaps unable fully to appreciate the "threat" of the risks involved. The patient may be all too eager to proceed when caution is clearly in order. Some candidates interviewed showed a limited grasp of the severity of the risks that indicated a considerable degree of denial. Lead surgeon William DeVries, during the days of the experiments with Jarvik 7, told of a woman who had called in to volunteer as a recipient of the artificial heart. She was not physically ill in any way, much less suffering from heart failure. Her motive was purely altruistic—she wanted to contribute to the advance of medicine! She was turned down, of course. She did not show sufficient self-interest to protect herself from known or foreseeable harm.

The problem is often one of a patient's limited imagination, not whether there are good faith efforts on the part of the physician to inform the patient. No intentional deception need be employed when a patient is convinced that the procedure will succeed with little more than personal discipline, an optimistic outlook, and attention to the regimen of drugs. The situation may be one in which false hopes on the part of the patient are met with the lure of optimistic hype on the part of the media and/or conveyed

[30] Mary Z. Pelias, "Duty to disclose in medical genetics: a legal perspective," *American Journal of Medical_Genetics* 39 (1991): 347–54.

unintentionally or subtly by the physician. How does one assess the impact of taking immunosuppressant drugs the rest of one's life? How does one imagine a life in which undesirable side effects of medication become routine: the hair loss, the gastrointestinal distress, the diarrhea, and the constancy of the regimen? It is a process in which patient denial is strong and motivation is partly stimulated by hype and the cheerleader effect.

A third factor is that the process of providing sufficient information and seeing to it that the patient understands what is said is time-consuming and frustrating to the specialist. Physicians understandably become impatient with a patient's slowness in grasping technical information. The temptation becomes strong to dispense with a truly informed consent and just procure a signature permitting the procedure.

Fourth, the physician is committed to professional and medical goals that are thought both desirable and justifiable. The aim and intention are to enroll a patient who seems a suitable candidate and is psychologically prepared to make the procedure succeed. After all, failure of the allograft would reflect poorly on the transplant team. The temptation becomes one in which the physician who is advocate for the patient subtly becomes the researcher who wants to make a medical breakthrough.[31] The patient is vulnerable to being manipulated into a decision that primarily favors the physician's interests. Robert Veatch has charged that informed consent may be another way of speaking of the process of getting the patient to acquiesce to physician wishes.[32] Both the patient and the physician are taking risks toward a more desirable future, but only the patient takes the life-threatening drugs.

In spite of its practical problems, informed consent is ethically required for any procedure, much less those in which avoidable risks are involved. The requirement is met when the patient has adequate understandings of the options available and their attendant risks to decide whether to accept an innovative procedure based upon one's personal values and life commitments. Those who insist the patient become an expert on the medical issues involved set a standard that is not only unrealistic, but ultimately self-defeating both for the patient and the advance of medicine.

---

[31] Paul D. Simmons, "Ethical considerations of artificial heart implantations." *Annals of Clinical and Laboratory Science* 16 (1986): 1, 2.

[32] Robert M. Veatch, "Abandoning informed consent," *Hastings Center Report* 25 (1995): 2, 5.

The Value of an Advocate

Matt Scot and other CTA recipients were required to name a "patient advocate" to assist with the process of interpreting the medical information, evaluating personal fears and hopes, reflecting on the pressures from media and health professionals, and/or raising questions that might have been missed. The transplant team believes this role of a "go-between" is a further assurance of reasonable objectivity on the part of a patient and the adequacy of information provided by physicians. Scott chose a longtime friend who is an attorney. Interestingly, the attorney was at first against the procedure. Scott finally persuaded him that going for the transplant was the best option for him.

The transplant team thus took pains to avoid two ugly faces of medical paternalism. One treats the patient as too ignorant to trust as a partner in new procedures. The other manipulates the patient to accept high-risk procedures in the interest of physician aggrandizement. The Louisville team took seriously the Kantian maxim that each person is to be treated with respect as an autonomous agent. The patient is not a means to an end. No more should be required of physicians than a good faith effort to provide the time and information necessary to deal with patient questions, fears, and hopes. In short, they are to deal with the patient as a person, not as an object of scientific interest.

A statement in the transplant team's requirements dealing with the type of candidate being sought did pose something of an ethical problem. "A young carpenter, mechanic, or surgeon who has lost the non-dominant hand whose state of health was so altered that the patient will accept the risks of hand transplantation" was listed among potential recipients or candidates for the procedure.[33] The question is whether too much was being promised when listing the possibility of a "surgeon, carpenter or mechanic" who might continue a career now jeopardized by an accidental loss of a hand. The competence of such professionals has a great deal to do with the dexterity and strength in one's hands. The statement seems to imply or promise too much, thus lending strength to the critics who charge that an overly optimistic hype was all too prominent in the process.

[33] Breidenbach,."The ethics of…." 17.

Personal Values and Quality of Life

The strongest objections to the allotransplant have come from those who feel that the risks to the patient are avoidable because the transplant is not necessary to sustain life. The statement by the Society for the Surgery of the Hand reflects this position. Unlike the situation with CTA, the argument goes, solid organ transplants are necessary to save a patient's life. The tradeoff of risks from side effects must be justified, in this approach, by the fact that it is a matter of life or death. As one surgeon put it, "I had rather have an immune system than a hand." Since the hand (larynx, knee, face) is not critical to "life," one should not be subjected even to a one percent chance of death for the sake of what amounts to a cosmetic procedure by this approach. In short, it is the argument concerning quality of life that surfaced so strongly but was not directly addressed.

Further, there are various considerations when examining the question of quality of life. If one would prefer a healthy immune system to a hand, what if the question deals with a face? Would a face be so much more important than a hand that the risk/benefit calculus might shift? The face is the way a person presents to others and the way we first perceive and relate to others. One man on the surgery ward had lost his face by injury complicated by diabetes. He now wore a prosthetic device attached to his glasses to disguise the fact that he had no face. When he removes his glasses, the space once occupied by his face now appears as a cavity. The response of most people when meeting him is a mixture of curiosity and shock. Would a transplanted face be of benefit or value to him, or would the possible benefit outweigh the risks from his perspective?

Quality-of-life concerns are considered under the concern for informed consent, but the objection used against the CTA seems to emphasize a doctrine that has lost considerable support in recent years as patients and the courts have struggled with values pertaining to "life." A reductionistic approach refers to life as the vital signs or the presence of life as opposed to death. A more comprehensive definition of "life," however, includes lifestyle issues and the values one considers necessary to make life meaningful or worth living, or that happen to be basic to the "pursuit of happiness" for this person. There are those who believe that quality of life is an even more important consideration than simply being alive.

Legislation pertaining to the right to refuse treatment, for instance, has favored quality-of-life considerations over a more reductionistic definition. Constitutionally, as the courts see it, the "right to life" is definitively related to "liberty and the pursuit of happiness." There are some circumstances under which one might prefer not to live. Ethically, quality-of-life concerns are related to the principle of autonomy or self-governance.[34] The decision either to forego treatment or to opt for risky procedures becomes a matter of highly personal considerations, reflecting beliefs and values that may differ substantially from person to person. As indicated above, informed consent is a matter both of understanding the options and their associated risks and of evaluating them in the light of one's moral values and religious commitments. A person's life is defined in terms of meanings and values, commitments and concerns, loyalties and beliefs, not just vital signs.

This insight goes to the heart of the decision regarding an allotransplant. In terms of quality of life, a patient might decide to run the risk of adverse side effects, including death, if the potential for restoring a lost voice, hand, or face is significant enough to make the risk worth taking. This legal principle was articulated by Judge Benjamin Cardozo of the New York Supreme Court: "Every human being of adult years and sound mind has a right to determine what shall be done with his own body."[35]

The rule is applicable to decisions ranging from refusal of treatment to abortion. It provides rather clear guidance for the issue of who decides in matters of accepting risky or life-threatening procedures. The desirability of an allotransplant may clearly be more important than the threat one feels from death. The patient is reflecting ground-of-meaning philosophical or religious commitments or beliefs that define life in a far more comprehensive way than simple vital signs. Whether or not to accept a risky procedure is, as the Supreme Court ruled in *Casey*, a matter of "the right to define one's own concept of existence, of meaning, of the universe, and of the mystery of human life." One's destiny "must be shaped to a large extent on (one's) own conception of spiritual imperatives and (one's) place in society."[36] In other words, patient autonomy prevails in the medical context regarding the acceptability of risks and benefits. The ethics of withholding what are perceived as medical goods in the name of "doing no harm" is to be seriously questioned. That is especially true since the nature of the "harm" to be avoided is so profoundly personal.

---

[34] Tom L. Beauchamp and James F. Childress, *Principles of Biomedical Ethics*, 5th ed. (New York: Oxford University Press, 2001) 63.

[35] Cited in *Cruzan v. Director, Missouri Department of Health*, 110 S. Ct. 2841 (1990).

[36] *Planned Parenthood of Southeastern Pennsylvania v. Casey*, 112 S. Ct. 2791 (1992) at 2807.

The problem that remains is whether the legal principle of self-determination makes medical procedures simply a matter of patient demand. The answer is no. It is a matter of patient "preferences"[37] rather than demands or mandates for just anything a physician might do to or for a patient. There are still procedures that cannot or should not be done, such as those addressed in the Nuremburg and Helsinki Codes, but where quality-of-life issues surface about procedures that have relatively low or otherwise acceptable risk factors, the burden of proof to show why they should not be done shifts to the physician or social critics.

If offered the option of transplant following the loss of an arm or leg from an accident and told the odds of rejection were less than 1 in 100, a reasonable person may well opt for the transplant. It is neither irrational nor compulsive for a patient to prefer to take the chance of recovering wholeness than to remain in a state of misery related to such a devastating injury. There is no firm line between one's personal well-being related to illness or disease and those based on an emotional or psychological sense of wholeness. Replacing a hand or face may be as important to one's sense of well-being and personal happiness as repairing a ruptured spleen or removing a diseased colon. Cosmetic effects may have powerful therapeutic benefits for the person, which must be weighed as seriously as the risks to life to which one may be subjected.

Conclusions

The future will undoubtedly see an exponential increase in the number of patients who desire an allotransplant for cosmetic enhancement. The paradigm is found in plastic surgery itself. What began as a technique for dealing with scarring from genetic anomalies, injuries, or disease has shifted into the science of enhancing bodily features. Now people are attempting to extend the appearance of youth and health by various nips and tucks. Liposuction and implants have become commonplace. Once anti-rejection techniques become more reliably therapeutic and less toxic, the demand for enhancement transplants will confront physicians with what are primarily ethical rather than technical questions. Fearing the future of misplaced priorities or questionable uses of a valuable therapy should not become an excuse for opposing its clearly therapeutic applications, however.

---

[37] Albert Jonsen , Mark Siegler, and William Winslade, *Clinical Ethics*, 5th ed. (Guilford, CT: McGraw-Hill, 2002) 93.

Other objections to the hand transplant have been based on such notions as (1) the immorality of allografts (or xenografts) because of the mixing of tissue between persons (or species), and (2) the problem of "playing God" on the part of surgeons who provide medical interventions to undo what God has done through nature or accident. That objection is apparently what prompted the courageous actions of surgeons in the fifteenth century to act in defiance of religious teachings and dare to transplant a leg, as the painting by Balbases depicts. The arguments raise further religious perspectives that reach beyond the scope of this chapter but are dealt with elsewhere in this book. Suffice it to say that such teachings have been refuted in both religious and philosophical circles. There is no intrinsic ethical or religious problem in providing an allograft for an amputee, any more than for providing a liver or kidney transplant.

The overriding commitment of the transplant team should remain that of providing competent and compassionate treatment for people who have suffered the traumatic loss of a limb or other significant body part. The goal of medicine is to provide the best medical care possible for the worst that human beings might suffer. The ethical grounds for the hand transplant fit solidly within the medical commitment to cure where possible but always to care.

The Louisville hand transplant team has rightly shown deference to public opinion and concern about the ethical issues involved, but no objection has surfaced to the CTA that is convincing to those who are equally committed to solid medical practice and the highest ethical standards. The reasons for proceeding with this significant but innovative procedure lie in the moral aims of medicine, which are correlated closely with the medical and personal needs of patients and the standards of ethics that guide the conduct of medical research.

Chapter 9

RELIGION, ETHICS, AND THE GREAT EMBRYO STEM-CELL DEBATE

> The one who believes in me will also do the works that I do; and greater works than these will be done, because I go to the Father.
>
> —John 14:12

> Whenever science is attacked on ideological grounds, its integrity and usefulness are threatened....moralistic dogma [must not] replace scientific judgment when the public's welfare is at stake.
>
> —Alan I. Leshner[1]

The debate over stem-cell research is an extension of the discussions regarding public policy and access to medical advances in areas such as in vitro fertilization (IVF), pre-implantation genetic diagnosis (PGD), and other techniques made possible by new genetic knowledge. Important breakthroughs have been made for infertile couples. Terms such as oocyte donation, gamete intrafallopian transfer (GIFT), embryo cryopreservation, and gestational surrogacy have become a standard part of the vocabulary in the world of reproductive technology. PGD now makes it possible to screen germinating life for major genetic burdens. Knowing more about the genetic basis of both disease and disability enlivens the hope that such knowledge will lead to interventions to prevent or correct genetic diseases.

Stem-cell research seems a major step in that direction and thus represents what may become a fundamental shift in the paradigm by which our bodies are treated and cured. By using the basic cells that make the development of all human organs and the neurological structures of the human body and brain, scientists now believe new organs can be produced

[1] Alan I. Leshner, "Don't Let Ideology Trump Science," *Science* 302 (28 November 2003): 1479.

to displace those which are damaged or diseased.[2] Science seems on the threshold of a remarkable new approach to disease—replace the broken, damaged, diseased, or deteriorated organ with one repaired or grown from the organ-producing cells of the body itself.[3]

Along the way, heated debates have emerged. Those who believe they are protecting important values relating to human life in its pre-embryonic beginnings seem pitted against those who are seeking to promote the health and well-being of patients in need of therapeutic medical interventions. The issues in the stem-cell debate are therefore not new. They were basic to the debate about abortion, genetic screening, and research related to pre-embryos from IVF techniques.

The debate involves not only values pertaining to complex medical research procedures, but how religious and moral values are related to the world of science and medicine. People motivated by powerful religious beliefs have entered the debate, defending doctrinal beliefs and commitments while attacking new methods and techniques in research medicine.

This chapter will attempt to shed some light on issues pertaining to religion and science, and their relation to the debate regarding stem-cell research. Examining the relation between theology, ethics, and public policy may help to set the stage for reflections and comments on aspects of the stem-cell debate.

Theological, Ethical Questions

Discussions of stem-cell research have raised certain theological and ethical issues that merit reflection and analysis. Certain preliminary observations seem necessary and appropriate before attempting a constructive statement on the relation between science and religion.

First, there is no firm line between theological problems and ethical perspectives, since each is related to the other.[4] Even so, a great deal of the problem in the religious debate in bioethics is that ethical issues are often isolated from their theological grounds or philosophical rationale. Both the moral argument and the theological presuppositions should be subject to

---

[2] E. Juengst & M. Fossel, "The Ethics of Embryonic Stem Cells—Now and Forever, cells Without End" JAMA, 284:24 (Dec. 27,. 2000) 3180.

[3] See "Stem Cell Surprise," *Science News*, Vol. 163, No. 9 (March 1, 2003) 131. Also see *Science News*, June 22, 2001, 390.

[4] See, J. M. Gustafson, *Ethics from a Theocentric Perspective: Vol. I, Theology and Ethics* (Chicago: University of Chicago Press, 1981) 87.

analysis and debate, of course. The current discussion is hardly a debate, however, since moral condemnations of stem-cell research are presented as if they are either self-evident or given by direct revelation. The result is to leave unexamined or unexplained the theological/philosophical underpinnings for what are presented strongly as absolute moral conclusions. The public is thus uninformed about the theological assumptions at stake and often misled as to the religious agenda(s) or perspectives involved.

Second, religion has no monopoly on moral insights or ethical actions. Religion provides no special knowledge, superior wisdom, or definitive rules to govern genetic studies or to settle thorny ethical dilemmas, even though some might prefer to think or claim otherwise. Certainly, some religious ethicists speak as if only *they* have an ultimate answer or a corner on the market of truth.[5]

The truth is that people of faith share the hopes and fears, the desires and concerns, of the general populace regarding the promises and pitfalls, the expectations and dangers, of innovative and ambitious scientific ventures in medicine. The primary guidance that can be offered by theologians is assistance in analyzing the questions posed by thoughtful people in terms of values, beliefs, and commitments that shape perspectives and actions. Theology can also shed light on the way(s) in which beliefs about God's providential actions in the world shape moral attitudes concerning human prerogatives as participants in the world of nature.

Theology certainly has an interest in stem-cell research and other genetic studies. Religious arguments regarding genetic studies reflect opinions about the ways in which the providence of God might be manifest and the ways in which what it means to be human might be at stake. Interests are quite different than absolute answers, however. There are good reasons to question the efforts of religious leaders to control science or prohibit certain lines of research. What is at issue may be prestige and power claimed by special religious interests, rather than the health and well-being of people or society.

We are all pilgrims in the venture of faith as we attempt to discern the will of God in the midst of human pain and suffering. God may reveal truth to whomever is available by faith and insight, whether secular scientist or concerned clergy. We need only be reminded that it has been hardly a decade since we observed the 400th anniversary of the church's condemnation of Galileo for saying that Earth was not the center of the

[5] Ibid., 30.

universe. Forced to recant his beliefs, Galileo continued his research and writings, but he was placed under house arrest until his death. Four hundred years later, and only after conducting a thirteen-year study, the Vatican gave final absolution to the astronomer. The Vatican statement admitted the error of the church's action, saying the condemnation was based on what Pope John Paul II called a "tragic mutual incomprehension."[6]

That historical episode is also a reminder that bad religion can get in the way of good science. In the midst of complex debates about serious science, we need to be reminded that God, not the church, is Lord of the universe.[7]

Third, theological beliefs do not yield obvious or immediate moral norms. The route from theology to genetics is often circuitous and always complex. Religious people typically affirm that God is Creator and believe that each person is created in the image of God. Beyond such general affirmations or beliefs, however, intense debates are generated when attempting to define the nature of God's creative activity and what it means to be made in God's image. Specific applications to issues such as those pertaining to protections that might be accorded to germinating life depend upon a pattern of related beliefs and their meanings for human action in the world. Moral norms must be developed that are consistent with but not particularly inherent in the theological affirmation itself.

Even so, ground-of-meaning beliefs establish the matrix for ethical reflection and deeply held moral convictions.[8] At stake in the debate about stem-cell research and genetic screening, for instance, are such theological issues as the relation of God to natural processes (where is God when a child is born with Tay Sachs or anencephaly?); the nature of being human; and divine creation and human stewardship. Questions for ethics are posed about the moral status of the pre-embryo, human procreative rights, and the moral meaning of research and therapeutic medicine. The debate is over

---

[6] *The Courier-Journal*, 1 November 1992, A8. James Carroll, in his book, *Constantine's Sword: The Church and the Jews* (New York: Houghton-Mifflin, 2001) 184 agreed with Hans Küng who says the "mutual incomprehension" comment implies there were errors on both sides. The mistake was only from one side, namely, the Church. Galileo was right.

[7] Gustafson, 89. Gustafson emphasizes the point that the reference point in theology is God, not people nor any human institution, including the church.

[8] See Glenn H. Stassen, "Critical Variables in Christian Social Ethics," in *Issues in Christian Ethics*, ed. Paul D. Simmons (Nashville: Broadman Press, 1980) 62f.

what might be called the art of "playing God."[9] Efforts to modify the randomness and chance at work in nature and thus sufficiently control its processes to reduce human suffering seem both necessary and problematic.

Issues of justice are also posed in terms of the access and distribution of medical knowledge.[10] The question of whether there should be federal funding for stem-cell research is related to questions of the political control of medicine and those belonging to "slippery slope" concerns. These problems are obviously interrelated, but cannot all be explored extensively in this chapter. Certain issues can be isolated for examination, however, and at least tentative ethical and theological perspectives can be offered.

The Pre-embryo as Person

The central theological issue posed by the stem-cell debate is the moral value of stem cells as particular instances of "human life." The clear preference on the part of those opposing research with pre-embryos seems to be that of using the terms "human being" or "human entity" when speaking of any type of living organism that is related genetically to people or human beings.[11] Linkages and equalities in value are found between any form of life associated with *homo sapiens*. Those who object to experimentation with stem cells, for instance, also object to research with or discarding pre-embryos or fertilized ova. Each of these is spoken of as a "human life," "human entity," or "human being" in the same sense as one would speak of a child. The assumption is that any instance of human life is of the same moral value that is usually attributed to a person.[12] Actually, those opposing stem-cell research rarely, if ever, speak of an embryo as a person, which seems a tacit admission that such a claim would be problematic.

That the stem cells or pre-embryos in question are "human" is hardly debatable. One can believe that they are "human" or "human life" without

---

[9] See Ted Peters, *Playing God? Genetic Determinism and Human Freedom*, 2nd ed. (New York and London: Routledge, 2003) 11. Peters discusses three meanings involved with the phrase, "playing God."

[10] See T. M. Garrett, H. W. Baillie, & R. M. Garrett, *Health Care Ethics: Principles and Problems*, 4th ed. (Prentice-Hall, 2001) ch. 4, "The Ethics of Distribution."

[11] See W. C. Kischer & D. N. Irving, *The Human Development Hoax: Time to Tell the Truth*. 1st ed. (Clinton Township, MI: Gold Leaf Press, 1995) who use the term "human being" from the time of fertilization.

[12] See H. Arkes, *Natural Rights and the Right to Choose* (Cambridge: University Press, 2002) 77.

believing they are entities that should be protected as if they were persons. The term "human" simply indicates the species to which they belong. There is no question that stem cells (fertilized ova, pre-embryos, zygotes, fetuses) are human if they are taken from or produced by *homo sapiens*, but are they persons? The answer is an unequivocal "no!" Whatever moral value or standing they hold is not the same as that of a person. A person is a far more complicated and developed creature than the simple structures or combinations of cells represented by an embryo, much less a stem cell. Those who use the terms interchangeably might reasonably be suspected of trying to obfuscate the issue. Their logic is a type of radical reductionism—reducing the entity that is actually valued to a single component of that complex entity so as to create a value transfer in the mind of the listener. [13]

The truth is that a stem cell or pre-embryo is human life, but it is not a person. The processes of gestation and cloning remind us of the enormous value of life at the embryonic level.[14] Germ line cells, for instance, are totipotent and thus might produce the entire human being. Embryonic stem cells, on the other hand, are pluripotent. These are early universal cells with the potential of forming almost any somatic cell in the human body. Such cells might allow the growth of transplantable organs in vitro, thus avoiding the problem of rejection.[15] The fact they have such potential does not make them of equal moral value to what they may become.

A fertilized ovum also has enormous potential. Given the proper uterine environment, nourishment, and genetic strength, the ovum might one day become a child. Having the potential, however, does not give an ovum the same worth morally or the same standing as that of a child. Most fertilized ova do not become persons, of course, since something like sixty to seventy-five percent of all conceptions end in miscarriage, according to embryologists.[16] Many pregnancies are never established since the fertilized ovum is discarded with the first menses. Others will be spontaneously aborted because of environmental (uterine) or genetic conditions inherent in

---

[13] See Paul D. Simmons, *Birth and Death: Bioethical Decision-Making* (Philadelphia: Westminster Press, 1983) 79–84.

[14] See J. Gearhart, "New potential for human embryonic stem cells," *Science*. 1998:282: 1061–1062. Also see R. A. Pedersen, "Embryonic stem cells for medicine," *Scientific American*. 1999, 280(4) 68–73.

[15] Juengst & Fossel, "The Ethics of Embryonic. . , ' 3180.

[16] Elizabeth Hall, "A Conversation with Clifford Grobstein," *Psychology Today*, Sept. 89, 43–46. Grobstein says that "only about one in three fertilized eggs make it to birth."

the developing embryo. Most fertilized ova will *probably not* become persons, contrary to claims made by some.[17] The plain fact is that a new DNA combination simply is not a person, however much some groups insist otherwise.

The argument by opponents of stem-cell research moves from contending that a stem cell or embryo is "human life" to the claim that its willful destruction is the moral equivalent of killing a person. The next step is to argue that one cannot experiment with such an entity without its permission, since we must have the informed consent of any human research subject. Further, the moral corollary is drawn that one cannot kill one person in order to benefit another or even many more should that be the outcome of an experimental research project. Thus, to sacrifice a pre-embryo (whether donated from IVF procedures or created for research purposes) is a moral evil, by this logic. At every point in the debate, the "life" at stake is taken to have the moral standing of and thus deserve the constitutional protections of those provided a person.

There are logical, moral, and biblical-theological reasons for rejecting the notion that a pre-embryo (stem cell, etc.) is a person, however. Logically, for instance, few people (if any) would deny the continuum from conception to birth, childhood and adulthood. That each person began as a fertilized ovum is an axiom of genetic and biological science, but that does not mean that every step on the continuum has the same value or constitutes the same entity. An acorn is not an oak tree, nor is an egg a chicken. An egg—even a fertilized egg—is still an egg and not a chicken. And one cannot sell acorns at the going price for an oak tree.

The genetic definition of personhood confuses potentialities with actualities. Certainly potentialities are important, but they do not have the same value as actualities. To blur the distinction is to abandon reason and common definitions important to scientific, social, and civil discourse.

The fallacy of the simplistic definition of personhood is also seen when the argument is reduced *ad absurdum*. Each body cell of a person contains that person's DNA, or genetic code. That is why, theoretically at least, persons may be cloned or duplicated using a differentiated cell from the adult body. The genetic definition of personhood necessarily but wrongly contends that each body cell is a human being, since each cell has the potential to become another person through cloning. The implications are

---

[17] J. T. Noonan, Jr. "An Almost Absolute Value in History," in *The Morality of Abortion: Legal and Historical Perspectives*, J. T. Noonan, ed. (Cambridge: Harvard University Press, 1970) 51–59.

also rather staggering when considering this line of reasoning for surgery or the removal of cancer cells from the body.

The fatal weakness of this argument, therefore, is its radical reductionism.[18] Equating "pre-embryo" (stem cell, fertilized ovum, or zygote) with "person" moves from a terribly complex entity (person) to an irreducible minimum (combination of chromosomes or potential for becoming a person). A pre-embryo is a cluster of cells that is hardly complex or developed enough to be considered or even function as a person. A person, or human being, has capacities of reflective choice, relational response, social experience, moral perception, and self-awareness. The zygote, much less a stem cell, has none of these. Both the person and the zygote have life, and both are "human" because they belong to the species *homo sapiens*. But a pre-embryo does not fully embody the qualities that pertain to personhood. A great deal of complex development and growth, both physiologically and neurologically, are necessary before the attributes of "person" are acquired.

Morally speaking, to claim that a pre-embryo is a human being is to introduce what philosopher Sissela Bok calls "a premature ultimate."[19] People have a near-absolute value in Western morality, but pre-embryos or fetuses do not; nor should they. They have value, but they are not of *equal* moral value with actual persons.

The Pre-embryo, Personhood, and Religious Doctrine

Of special concern are those proposals regarding the pre-embryo as person that rest on abstract metaphysical speculation. The notion that a pre-embryo is a person most often claims that it has been infused with a "rational soul" at the moment of fertilization.[20]

The language of "soul" is extremely problematic for ordinary discourse. Not everyone agrees that there is such a thing as a soul, and religious groups disagree strongly as to the nature of soul,[21] but very few religious groups believe in soul infusion, as if an entity from outer space has invaded and captured a body at the moment of conception. Further, the notion of soul is ultimately unprovable and the debate irresolvable. Hindus believe cows have souls, but Christians do not. Christians attempting to

---

[18] Simmons, 83.

[19] S. Bok, "Who Shall Count as a Human Being?" in *Abortion: Pro and Con*, ed. R. C. Perkins (Cambridge: Schenkman Publishing Co, 1974) 91.

[20] Noonan, 53.

[21] David A. Jones, *The Soul of the Embryo* (London: Continuum, 2004) 93.

overturn *Roe v. Wade* argue that a zygote has been infused with a soul. Neither idea should be the foundation of law in a pluralistic society. The American Constitution attempts to protect minority religious beliefs from the imposition of alien/odious religious dogma by dominant groups. The American social contract calls for the separation of church and state. The notion of soul, whether defining the moral value of fetuses or cows, is rooted in abstract metaphysical speculation that defies resolution.

The issue for public policy is a definition of personhood that is appropriate in and for a pluralistic society. The first mandate of the Bill of Rights guarantees that "Congress shall make no law respecting an establishment of religion," which means, among other things, that laws will not be based upon abstract metaphysical speculation. Metaphysics and the fine points of theological reflection are far more appropriate for the deliberations of religious councils, not the halls of Congress or legislative groups. Law or public policy should be fashioned through democratic processes in which every perspective is subject to critical analysis. Secularists as well as people of faith must be able to participate in the discussion and have their particular religious interests protected in the final outcome.

The Genius of Roe. Part of the genius of *Roe v. Wade* (and *Casey*) was the distinction between actual, indisputable persons and beliefs about personhood that relied upon abstract metaphysical or theological appeals. For purposes of public policy, the court has rightly insisted on a definition that corresponds to commonsense and scientific understandings of biological and neurological development that are necessary and indispensable to being a person. In matters of religious beliefs, the various traditions may formulate a perspective representing even idiosyncratic notions. People are free to believe as they will, but just any religious belief cannot be imposed on all people in a pluralistic society. Public policy must be based on commonsense reasons consistent with the rule of law and not entangling matters of state with those of religion.

In *Roe v. Wade*, the court focused on viability as that stage in fetal development that one might be recognized and protected at law. Prior to viability, the fetus simply is not sufficiently developed to speak meaningfully of it as an independent being deserving and requiring the full protection of the law, that is, as a person. The notion of viability correlates biological maturation with personal identity in a way that can be recognized and accepted by reasonable people. It violates no premise of logic to provide certain legal protections for a viable fetus. Most religious groups agree with this assumption. At that stage of gestation, the fetus is far more like a newborn infant than it is a fertilized ovum.

The same cannot be said for efforts to establish moral and legal protections for just any form of human life as if it should have equivalent value to persons in morals and law. Defenders of the reductionistic approach often claim that a fertilized ovum is a person whether other people agree or not! In other words, no matter what ordinary logic might indicate, the opinion of the theologian or moralist is actually the truth. Those who disagree are thought misguided, uninformed, or willfully ignorant. In spite of its problematic theoretical basis, there is a strong effort to fashion civil law and public policy based on the contention that the strict moral rule against research or experimentation with any form of "human life" should be implemented by civil law.

Freedom of Religion. That many people genuinely and sincerely believe that stem cells or a pre-embryo should have legal protections is by now well established. The First Amendment allows people to believe as they will as a matter of conscience or religious belief, that is, as a freedom *of* religion,[22] but as a definition of personhood for public policy in a pluralistic society, the human life-as-person rationale is untenable in the extreme. One cannot appeal to "soul infusion" or other abstract metaphysical notions and claim to be respecting the difference between church and state, or faith and politics. John Rawls has it right in saying that definitions for public policy must be supported by ordinary observation and modes of thought that are generally recognized as correct.[23]

Ethics and the Genetic Prospect

Support for stem-cell research has two major ethical groundings. The first is in the moral mandate that underlies and motivates medical science. The ethical issue of *not* proceeding with such research is that we neglect and ignore actual persons in favor of conceding status to "potential" persons. The pre-embryo or stem cell regarded as of equal worth with persons is actually a "metaphysical abstraction." To love a stem cell is hardly the moral equivalent of loving a person. We have a moral duty to care for people.

---

[22] See *Cantwell v. Connecticut*, 60 S. Ct. 900 (1940). As Justice Owen J. Roberts wrote: "[Freedom of religion] embraces two concepts—freedom to believe and freedom to act. The first is absolute but, in the nature of things, the second cannot be." Ibid. at 903.

[23] John Rawls, *A Theory of Justice* (Cambridge, MA: Harvard University Press, 1971) 213.

Christians would put it in terms of the moral requirements of agape: loving the neighbor as oneself.

The Medical Mandate. Stem-cell research seems a medical mandate based on the demand to love persons by seeking their health and wholeness. Using stem cells as a way to perfect techniques to assist persons to become whole or regain their health that has been damaged by injury or disease seems a corollary to the love commandment. The moral concern for those afflicted with dread disease, for those not yet conceived, or the unborn who will be handicapped, and a desire to see stronger, not weaker, persons in the future, seems to mandate a response that is itself ethical, namely, to do all that is possible to prevent or reduce human suffering and limitation from disease and deformity.

Both good morals and good medicine require that we do what we can to prevent human suffering and cure diseases that severely limit persons and/or condemn them to a premature death. People are also custodians of the future. Some anomalies affect the children of each successive generation more severely than those who have gone before. Myotonic Dystrophy, for instance, is a degenerative muscle disorder that causes victims to be able to grip but not let go. The person also suffers from cataracts, abnormal heartbeat, diabetes, and mental retardation. And the symptoms get more severe with each generation. There may be fifty repeats of the CTG (cytosine, thymine, and guanine) sequence in the first afflicted generation, and as many as 2,000 repeats in their children and grandchildren. The more repeats, the worse the symptoms. A central principle of medical ethics is that physicians are to do no harm. Non-maleficence (do no harm) also requires that physicians are to *prevent* harm where possible. If devastating replications of CTG can be prevented, they should be. If stem-cell research can lead to such cures, it should be done.

Stem-cell research is a moral response to tragedy and misfortune in the human family. Disease and deformities have tragic consequences for the person affected, for their family, and for society at large. At the personal or individual level, we are touched by the tragedy and unpredictability of viral or bacterial infections and genetic mutations. We are confronted with the randomness and chance to which we and our children are subject. The womb is a dangerous place.

The science of genetics is helping to name the problem and marshal resources to do something about it. The problem is the human experience of evil and suffering, which takes the form of genetic diseases that maim and cripple, hideously disfigure, and then kill painfully and prematurely.

Personhood as Proactive Agentry. The second major ethical grounding for stem cell research may be found in a philosophical or religious understanding of human personhood. An important dimension of being human is to resist passivity in the face of threat to personal well-being, whether individual or corporate. As moral agents, people are to be participants in shaping the world and its future, not passive spectators of a tragic pattern in nature to which they are simply to be hapless, helpless victims. Harm is to be avoided and prevented whether in history or from the deleterious effects of random genetic mutations.

The presence of disease in nature may be viewed as a challenge from God. The divine "command" is that people pursue knowledge until ignorance no longer prevails, subjecting people to the futility of unnecessary and tragic suffering. The Apostle Paul seems to suggest that a great deal of the pain and distress of creation is because of the ignorance and negligence of people who have not found cures that are "there" to be discovered. He declares that "the groanings of the present world...are awaiting the disclosure of the children of God" (Rom 8:22–23). Paul seems to be saying that when "the children of God" act responsibly, creation will not suffer as extensively. God's children are those who accept the burden, challenge, and responsibility to gain new knowledge in order to bring about cures or corrections.

Such knowledge is an extension of human personality into the realm of science and technique. People are toolmakers and designers precisely because they image the powers of God. Technical knowledge is a basic ingredient in being human and indispensable to the task and goals of human stewardship.

Perils and the Promise. To be sure, stem-cell research is a venture holding both perils and promises. Human sin and dignity are related in the human psyche. Science is not a purely altruistic venture. Ambition, greed, power, and prestige are all dynamics in medical research, but they do not undermine the moral basis of research any more than the world of medicine itself should be dismissed as morally unworthy because some physicians have been found incompetent or morally unworthy.

Our quest for health and the well-being of the genetic future drives us to find cures and prevent illness and handicaps as far as is possible, but the human capacity for evil and the harmful effects that result even from most beneficent actions mandate guidelines for caution and restraint. Both our fears and our hopes are justifiable and should be taken seriously.

The Number and Causes of Abortion. Stem cell research also offers the promise of reducing the number and causes of abortion. A common cause of pregnancy interruption, whether intentional or spontaneous, is related to radical fetal deformity. Tay Sachs, anencephaly, Down syndrome, Lesch Nyhan, and a host of other genetic defects can now be discovered by tests for genetic markers in the very early stages of pregnancy. Such knowledge early in gestation now makes abortion an alternative to childbirth for numbers of couples.

On the other hand, there are certain types of genetic screening that make choosing abortion for reasons of genetic burden unnecessary. Pre-implantation genetic diagnosis (PGD)[24] of the IVF pre-embryo, for instance, eliminates the need for abortion as an option. The decision can be made in the laboratory before attachment to the womb as to which pre-embryo is affected and which are normal. Those found without the lethal genetic defect can be returned to the womb in order to establish a pregnancy. The genetic health of the child-to-be can thus be assured while fulfilling the woman's desire to become pregnant. Prospective parents can be virtually certain that the child-to-be will not be burdened because of genetic anomalies.

The diagnosis for conditions like Tay Sachs, Lesch Nyhan, cystic fibrosis, and hemophilia looks for a specific mutation. The diagnostic technique is 100 percent accurate. Those pre-embryos that lack the genetic defect can be used for implantation. Three is the optimal number to return to the uterus in order reasonably to establish a pregnancy. Returning a larger number runs the risk of a multiple pregnancy that might necessitate selective reduction of the number of embryos. Those pre-embryos that carry the genetic defect can either be discarded or used for further research. It would be morally questionable either to return them to the womb or to offer them for possible adoption, hoping that God or chance would see to it that a healthy child would be born, in spite of the known genetic burden.

Moral Values and the Pre-embryo

How, then, shall the pre-embryo that is the object of exploration, testing, and perhaps even modification, or as the source of stem cells, be viewed morally?

---

[24] R. J. Tasca & M. E. McClure, "The Emerging Technology and Application of Pre-implantation Genetic Diagnosis," *Journal of Law, Medicine and Ethics*, 26 (1990) 7–16.

The Vatican Response. The strongest objections to such procedures come from those who regard a pre-embryo as a person, as noted above. IVF procedures are condemned absolutely by such standards since it is argued that persons are being created and experimented upon without their consent. The moral gravity is compounded when fertilized ova are discarded or otherwise disposed of, which is regarded as killing persons or the murder of human beings. Pope John Paul II lectured President Bush during the President's visit to Italy in July 2001 as to what the Pope regarded as the moral evil of stem cell research. The Pope maintained that his attitude toward stem cells was consistent with the Vatican's condemnation of any procedure having to do with embryos.

The Vatican's Congregation for the Doctrine of the Faith had also issued strong condemnations of most forms of treatments for infertility such as IVF and embryo transfer. The document called for the recognition of "the dignity of the person" from the very "moment of conception."[25] No embryo stored by an IVF clinic is to be discarded—not even those affected by the most hideous genetic flaws!

The document does say that therapeutic treatments for flawed pre-embryos would be morally permitted. The catch–22 in that statement is that no research on a pre-embryo is morally permitted unless there is nearly absolute certainty that no harm will be done. That stricture is tantamount to forbidding research that might result in therapeutic treatments. Science cannot develop therapies that are known to be helpful without research and experimentation and the resulting losses that are basic to the trial-and-error method of medical research.

The Constitution and the Pre-embryo. Thoughtful people are grateful that the principle has now been established at law that a pre-embryo does not have the standing of a person in the United States or most any country in the Western world. The Supreme Court's decision in *Casey* (1992) effectively struck down laws that had been enacted in Illinois, Missouri, and Louisiana that spoke of "the right to life of all persons from the moment of conception."[26] IVF procedures and most forms of genetic screening would have been illegal in such states. By implication, the ban would also be extended to stem cell research. There would be a troublesome moral irony

---

[25] Cardinal Ratzinger and the Congregation for the Doctrine of the Faith, *Instruction on Respect for Human Life in its Origin and on the Dignity of Procreation: Replies to Certain Questions of the Day*, (Rome: Vatican Press, 1987) Pt. I, 1. [See Appendix V.]

[26] *Planned Parenthood of Southeastern Pennsylvania v. Casey*, 947 F. 2d 682 (3d Cir. 1991) 112.S. Ct. (1992).

in public policy that preferred the rights of a stem cell or a pre-embryo to those of real or actual persons.

The Value of the Pre-embryo. Denying that a pre-embryo is a person does not mean that it has no value. It is precisely that it is *potentially* a child, *potentially* a person, that it has enormous value to those women or couples who desperately want to have a child, or are searching for new and more effective cures for disease and disability that have shown up in children already born or pregnancies lost.

Furthermore, the pre-embryo can provide stem cells, which are the source of life and health-giving material. Scientists may use stem cells to develop vital therapies for people who suffer from disease or injury. Stem cells are unique in that they offer prospects for a range of treatments and cures that simply have no other parallels in medical science. Surgeries can correct defects but cannot cure a heart that has deteriorated past the point of recovery. Transplants can be done, but there is an enormous scarcity of available organs, and the recipient must use anti-rejection drugs, which have undesirable side effects, for the rest of one's life. Procedures using stem cells offer the potential of growing new and healthy organs to replace those that are diseased and atrophied. It promises to be medical therapy like no other.

The Ethics of Research with Chosen Pre-embryos

Even so, the value of the pre-embryo does not depend upon regarding it as a person. Rather, the moral value of the pre-embryo rests in the fact that it offers material for further knowledge of the basic stuff of genetic inheritance. The pre-embryo is chosen for a special role in bringing about personal health and well-being. It is valued for its potential to enable scientists further to assure the benefits of health to our children and theirs. Research is as much a moral mandate as is therapy. It seems far better to manipulate the building blocks of bodies and brains than to try to correct or treat severe illness after extensive damage is done. God has given us the grand opportunity to discover how to prevent, as well as treat or cure, the illnesses that plague humankind.

The pre-embryo will thus be treated with due regard or respect without lapsing into a posture of reverence for its "life" or believing that it is in some sense "sacred." Researchers are to respect people, and to revere God, and the two should not be confused, as Karl Barth noted. The embryo is neither a person nor God. Nor is it just raw material; it is a gift from God that may aid the wholeness or health of human bodily existence.

The givens of nature are not the creations of the scientist. Research scientists do not create out of nothing. They work in a world that is not of their making. Science is to that extent like theology—it is faith seeking understanding. The calling or mission of science is better to understand the forces that affect us so they can be manipulated for good rather than ill. In exploring the cells of a pre-embryo, the scientist views not only God's creation, but the promise of the human future. As long as research is done in the context of reverence for God and respect for the future of humanity, it is essentially a moral enterprise.

Learning to prevent genetic illness is as moral as Jenner's learning to prevent smallpox. Stem cell research is not morally unlike vaccinations to prevent disease or pre-implantation genetic screening to prevent the birth of children with radical birth defects. The bacteria that bring illness and death have a certain moral symmetry to the deleterious genes that infect our DNA. Bacteria and bad genes are different sorts of invasion of the biological processes at work, but they both affect our health and well-being. The genetic influence is perhaps more sophisticated, complicated, and insidious, but the end result is similar—disfigurement, disease, and premature death. As long as these afflict the body of humanity and threaten the genetic future of people, science has a moral mandate to pursue medical knowledge through genetic research. That task will include stem cell research and the use of pre-embryos. We cannot choose either passivity or ignorance toward disease and affliction and claim to be moral.

Summary and Conclusions

Stem cell research rests upon the moral mandate that drives all medical science, namely, the search for better and more effective medical therapies.

Pre-embryos are not persons in any ordinary sense of the term, nor should they be protected under the law as if they were persons. Even so, they have a dignity in that they are chosen, dedicated to particular and unique purposes related to human health and well-being. Those used in research are either the result of donations for experimentation by couples treated for infertility or are created in the laboratory for research purposes. In either case, the pre-embryo is respected for the unique role it may play in health and healing.

Stem cell research has significant ethical values. First, it may practically resolve the abortion dilemma for genetic anomalies. Second, it provides relative assurance of the birth of a healthy child. If lethal anomalies can be corrected with stem cell therapy, it will reduce the suffering to which so many people are condemned by disease and deformity. Third, it reduces the health cost burden to families and to society by reducing the numbers of persons with incurable and perhaps lethal illnesses. Finally, it contributes to the general health of the human future.

Federal support for stem cell research would also be entirely consistent with and supported by Supreme Court decisions since *Roe v. Wade*. The legal standing of the pre-embryo is now relatively settled in US law. Affirming stem cell research will assure that contradictory approaches between the executive and the judiciary will not lead to further constitutional clashes.

A positive response to stem cell research is also consistent with the First Amendment guarantees regarding the separation of church and state. Narrowly construed religious dogma that defies ordinary logic is not to be the basis for law in America. All people should be free to believe as they will, but no one should be coerced to live with policies based upon abstract metaphysical speculation that is appropriate to theologians but not to legislators.

For all these reasons and more, US public policy should support stem cell research. The President has the grand opportunity to provide leadership that encourages science and refuses to bow to powerful coalitions that pursue private or religious interests at the expense of actual people and the health and well-being of society.

Further, federal funding of stem-cell research will provide an avenue of oversight that will be impossible if research is left to private funding. The research is already taking place. Major drug companies are competing for the patent rights and marketing advantages that come with initial breakthroughs and developments of therapeutic interventions or research procedures. The public interest is best represented by government oversight and structures of accountability, in spite of their limitations.

A positive public policy toward stem-cell research will recognize and institutionalize the broader support for science that is generally given by world religions. Good religion and moral values recognize that such research is made possible by moral agents who are exercising powers that are uniquely human and for purposes that serve human well-being. We are no longer simply helpless, hapless victims of circumstances beyond our control, but are pursuing the human hope for a more perfect and healthy future.

Chapter 10

FAITH, ETHICS, AND ABORTION: THINKING ABOUT THE UNTHINKABLE

In the tense days of the Cold War between the Soviet Union and the United States, the prospect of nuclear war was a source of enormous anxiety and stress. The official policy of the superpowers was called MAD (mutually assured destruction). The horrors of war were thought so terrible that only the deranged would use a nuclear weapon. Herman Kahn, a physicist with MIT, dared to question that assumption in a little book called *Thinking about the Unthinkable*.[1] His argument was that people had better take seriously the prospect of nuclear war, not be lulled into the false security of thinking it would never happen because it was simply too horrible to imagine.

Kahn's warning seems appropriate for thinking about, analyzing, and responding to the abortion debate. Few issues have generated such intensity of feelings and acrimony in rhetoric. The arguments illustrate a central contention of this book, however, that religious faith, assumptions about scientific teachings, and concerns for supportable public policy are all at stake in the issue of abortion. The issue is of importance as a matter of morality in reproductive choice and in terms of politics and legal options.

This chapter will outline and examine the oppositional perspectives in terms of what each finds unthinkable and thus unacceptable. Both sides aim to affect public policy—one aims to ban the procedure, the other to protect its legal availability. My own position is that it is unthinkable that abortion should be made illegal in the United States. I will outline those concerns both by giving what I think are solid reasons in favor of choice and by showing what seems to be the flagrant errors of those trying to ban abortion. The anti-choice crusade is targeting liberties pertaining to choice that threaten the very nature of American life and religious understandings of responsible freedom before God. I do not advocate abortion; only enemies

---

[1] Herman Kahn, *Thinking about the Unthinkable* (New York: Simon and Schuster, 1984).

of choice portray those who support choice as pro-abortion. I advocate a legal climate in which women and families may consider their options and make tough decisions in light of their own religious commitments.

Abortion as the Unthinkable

The story of Rabbi Shira Stern captures the oppositional perspectives in the debate. She, along with her husband, Rabbi Donald Weber, decided to terminate a pregnancy after sonograms showed the baby was anencephalic (no brain) and severely deformed. Their decision to abort had the full support of their religious tradition and the community of which they are a part. From her hospital bed, she heard a strongly worded, televised anti-abortion message by then-President Ronald Reagan. He declared that abortion is "this nation's number one moral problem" and promised to press for legislation to outlaw "the murder of unborn children."

Reagan articulated what became the mantra of the anti-abortion crusade in America. Rabbi Stern captured the pathos and tragedy that goes into the call to keep abortion legal and to recognize the circumstances that women face that often lead to abortion. Abortion is not always wrong, as Reagan would have us believe, but such choices are most always the result of difficult and complex circumstances.

Critical Variables in Conflicting Opinions

The powerful emotions at work have most often displaced any interest in looking critically at the variables that underlie the conflicting opinions. A great deal of heat has been generated, but light has been obscured, scarce, or absent.

The place to start is to recognize that religion seems the most important motivating factor on both the drive to ban abortion and the effort to preserve abortion as a legal option. It is not that one side is religious and the other is not. The debate in America is primarily among people who claim a religious or even biblical basis for their position. Where they differ is in their beliefs regarding the moral status of gestational life, understandings of personhood, the role of law in regulating reproductive choice, and the freedom of the individual believer before God.

The meaning of religious authority is also at issue. Anti-choice groups support external authority centered in religious structures or legal mandates. Those emphasizing choice believe the individual must answer directly to God for responsible living and reproductive choices. From this perspective, the Grand Inquisitor has no place in religious life since obedience is directly to God, not an ecclesiastical religious figure or a legislative mandate. The law does not settle the moral question, but it may greatly complicate the difficult choices facing women and their families.

The type of society envisioned for the future also differs in substantive ways. Anti-choice groups prefer a government that is dominated by religious authorities. At least three approaches to the relation of church to state can be found in the coalition against choice. Some prefer the pattern of the Puritans at Salem, with its influence of Calvin's experiment in theocracy at Geneva, Switzerland; still others follow the vision of the Roman Empire, where the Pope and clerics ruled by using government power to enforce doctrinal and moral teachings; and the third still yearns for the English effort to impose laws requiring religious and doctrinal conformity. All three have in common a type of theocratic relation between church and state. The basic assumption is that religion supplies the moral rules, and the legislators impose the legal sanctions.

Those who champion choice envision a more open society, fearing the excesses of church-state entanglement. The writings of those who fashioned the American Constitution showed their determination that the oppressive, judgmental, and destructive eras of history in which religion wielded such a heavy hand over all citizens would not be repeated in this country. Democratic governance may not be perfect, but it is far superior to the structures of injustice perpetrated by religious groups who gain political power. Protecting individual liberties is vital to preserving a democratic and egalitarian society. The values associated with liberties that are religious, personal, and political are worthy of sacrificial pursuit and determined preservation.

Both groups embrace a certain vision of the good society. Those aiming to ban abortion believe the purpose of law is to make people good, which means *on their terms*. They are not motivated by a vision of a *just* society that champions liberties instead of conformity to externally imposed rules in the name of religious morality. Anti-abortion forces are targeting a number of issues for legal control and/or prohibition. They are authoritarian and oriented toward dominating or eliminating oppositional perspectives on religious and social issues. Those who love and are willing to make sacrifices for freedom find it unthinkable that political power would become the enforcement arm of religious intolerance in America.

The anti-choice crusade has also manipulated the science of embryology to fit its ideological claims. The claim that a fertilized ovum should be protected as "innocent human life" needs the careful analysis of evidence-based science, but anti-abortion forces want to repeat the mantra of "murdering innocent human life" so as to create a climate of guilt and shame to control the popular mind and medical practice. A conceptus is not a person and should not have constitutional protections, as the Supreme Court has repeatedly said, but those determined to outlaw abortion are intolerant of informed opinions, preferring dogmatic ideology to democratic procedures.

Scientific and medical opinion regarding the status of embryonic life has not shifted in the last three decades. New technologies enable viewing the fetus during various stages of development *in utero*, but there are no instruments to measure the presence of personhood. People agree about the gestational facts, but disagree about their value.

There are four stages of gestational development. The *zygote* is a fertilized ovum; when attached to the uterus, during which time cell division occurs, it is a *blastocyst*; the *embryo* is the two- to eight-week stage, during which vital organs and blood vessels begin to develop; and the *fetus* is from eight weeks to birth, during which biological and neurological maturation continue.

Embryologist Clifford Grobstein points out that human emotions begin to attach to the developing embryo during gestation. The woman or the parents may declare it a person and value it as if it were already a member of the family.[2] Either as a religious doctrine or as a personal attachment, many are persuaded that an embryo is a person. That is as it should be since these emotional ties form the commitments to nourish, protect, and cherish the embryo toward personhood and membership in the human family.

We can all agree that each person began with conception and implantation, but that does not mean that every step along the continuum has the same value. There are important differences between facts and values. Science provides the data of biological development; religion and personal values provide our sense of morality and the worth we attach to various entities. Functional physiological and neurological developments are required for one to become a person, but there are stages during gestation in which there is simply inadequate development to constitute an entity that should be accorded full legal or even religious status as a person. Spontaneous abortions are rarely given funerals or treated as subjects by religious groups, which underscores the point.

---

[2] See the interview by Elizabeth Hall, "A Conversation with Clifford Grobstein," *Psychology Today*, September, 1989, 43–46.

As Grobstein points out, full individuality emerges in stages over time. A person is a complex creature that has moved from genetic individuality (DNA) through functional, behavioral, psychic, and social dimensions. A genetic formula is not the equivalent of a person, nor is it "a person with potential" as the anti-abortion crusade would have us believe. Repeating a misleading ideological formula a thousand times will not make it the truth, though millions have apparently accepted the idea.

Physicians are divided as to whether the woman should have the final say about abortion. The intimidating tactics of the anti-choice movement have radically reduced the number of physicians willing to do abortions, however. Surveys show that many physicians who are sympathetic with the need for abortion services are unwilling to run the risk of being targeted by ugly demonstrations or being wounded or killed. Drs. David Gunn and Bernard Sleppian are among those physicians who dared to provide services in spite of threats against their life. Tragically, they were assassinated.[3]

The more the courts affirm the legal availability of abortion on constitutional grounds, the more radical the opposition becomes. Anti-abortion forces say they find it unthinkable that germinating life might be expendable, and the fact that the woman's life should be protected (perhaps at the expense of the fetus) throughout the nine months of gestation infuriates them even further. The radical nature of this crusade has rather frightening implications for women's health and the shape of public policy.

The Nature and Goals of the Anti-abortion Crusade

Opposition to legal abortion is led by the "right to life" movement, which unites orthodox Roman Catholics with ultraconservative fundamentalist/evangelicals. Their strategies include delaying tactics, restrictions on availability and licensing of abortion clinics, and intimidation/persuasion of women contemplating abortion. Activists may block clinic entrances, lecture or shout at the woman, and organize demonstrations and pray-ins. The clerics involved have been known to threaten pro-choice politicians with excommunication or denial of Communion, as was done in the 2004 elections.

Not a few commentators have raised questions about the morality of means often employed by the coalition on the right in its pursuit of its

[3] See, Eyal Press, "My Father's Abortion War," *New York Times Magazine*, Sun., Jan. 22, 2006, Sec. 6, 57–61. Since the late 1970's seven abortion providers have been murdered, 209 clinics bombed or burned, and hundreds of death threats made against clinic staff or physicians, according to the National Abortion Federation.

social, political, and religious objectives. The misrepresentation of the theological and moral positions of its opponents, slanderous accusations, deceit, and the use of fabricated stories to illustrate their points or to sway emotions and opinions have all been documented. They seem willing to "lie, cheat, and steal" in order to accomplish their social and religious goals. They have strongly embraced the politics of prevarication. As a businessman in Houston, Texas, said, "Never in six decades in the business community have I encountered motives more questionable, manipulations more deceptive or methods more merciless, than by men claiming to defend the Bible, who seemingly have voiced obligations to live by God's Word."[4]

The reason for what critics see as a moral blindness or inconsistency may be explained by the adoption by the religious right of what Roland Bainton called "a Crusade ethic."[5] Four characteristics of this style of moral reasoning were described by Bainton. First, the war has a holy cause; it is not about justice. That is, the motive and goals are religious, not secular, in nature. Second, God is said to fight on behalf of the Crusaders and against their enemy. Third, the Crusaders claim they are godly and righteous, while the enemy is ungodly and unrighteous. Fourth, the war is prosecuted unsparingly. There are no manners or rules of civility that govern the nature and degree of the conflict or the choice of instruments or strategies involved.

Whether in the Inquisition or the Puritan theocracy, cruel and unusual punishments, including public ridicule, imprisonment, beatings, and other denials of human dignity, were employed in the name of religious piety and zeal. The actions took place under the guise of moral uprightness and with the protection and/or collusion of government and religious authorities.

Fundamentalist evangelicals, like the Puritans, are fierce opponents in religious and political battles. They have a macho-Rambo-style conflict. The tougher-than-you-are approach fits squarely with their style of leadership in foreign and domestic matters and with the "masculine" image their leaders cultivate.

Anthropologist James Prescott says there is a consistent pattern of intolerance toward others and a proclivity toward violence among anti-choice advocates. In most all cultures, he finds a predictable correlation between being anti-choice and supporting such institutions as slavery,

---

[4] John Baugh as quoted in *SBC Today*, June, 1989, 5.

[5] R. Bainton, *Christian Attitudes Toward War and Peace* (New York: Abingdon Press, 1980) 66.

killing, torture, mutilation of prisoners of war, harshness toward women and children, and capital punishment.[6]

This manner of "making war" is given religious underpinnings since God's power and wrath are construed in terms of cruel vengeance or stern vindictiveness. In orthodox or fundamentalist theology, God's justice requires punishment for wrongdoing. Stories of ethnic cleansing or genocide in the Hebrew Bible are thought literally mandated by God. God is often portrayed as the cause of suffering in human life when tragedies strike. Whether genetic deformity, pregnancy by rape or incest, or tragic, sudden death by stroke or auto accident, God is thought to be the cause of the tragedy. The reason for the harsh action is found in human sin or lack of piety.[7]

Some evangelicals believe in a God of wrath and an ethic of doing violence in the name of fidelity and service to God. The Spanish Inquisition had dissenters (Protestants, Jews, atheists) killed in the name of godly righteousness. It was actually thought that killing heretics was necessary to save them from the damnation of hell to which wrong beliefs in doctrine and morals would surely lead. Torture, imprisonment, and death were thought merciful favors in order to bring heretics to repentance.

Few movements are so to be feared or more fervently resisted than those of religion and politics joined in a crusade to promote decency and righteousness. Any ungodliness can be justified in the name of zeal for truth and morality! When religious zealotry and ideological passions combine with political power, the first cousin of fascism emerges that cares neither for God nor people while pretending it serves the interests and will of both. That ugly alliance can be found in fundamentalist Islam, as in Iran under the Ayatollah, or al-Qaeda under Osama bin Laden. And it can be found in Judaism and in militant Christianity. Nazi Germany exploited religious bigotry and mindless patriotism among Christians and used it to justify pogroms against Jews, Gypsies, homosexuals, and others.

Truth is one of the first casualties in a crusade where winning means everything. The question of truth is central, for instance, when evaluating the ethics of crisis pregnancy centers, which have multiplied by the dozens near clinics providing abortion services. When a woman mistakenly goes to

---

[6] James W. Prescott, "Personality Profiles of 'Pro-Choice' and 'Anti-Choice' Individuals and Cultures," in *Abortion Rights and Fetal 'Personhood*, eds. Edd Doerr and James W. Prescott (Long Beach, CA: Centerline Press, 1989) 115 ff.

[7] See James T. Draper, *The Conscience of a Nation* (Nashville: Broadman, 1983) 87–88.

the counseling center, she is not given all-option counseling or compassionate guidance. She is shown pictures and given materials on fetal development with a strongly worded narration in an effort to dissuade her from having an abortion. Understandably, women are often shamed or cajoled into changing their minds. If the ethical issue is raised about using deception to lure the woman into the clinic and then using tactics that are humiliating and shameful to the point of leaving her in tears, the answer is some variation of the theory that a good end justifies what might otherwise be a questionable means. As one physician said, "Anything I do to stop a woman from having an abortion is justified."

The Bush administration has poured millions of dollars into these centers, usually under the guise of faith-based initiatives. At the same time, it has de-funded and created barriers to the work of groups like Planned Parenthood, which is committed to decreasing the incidence of abortions by providing accurate information.[8] The "information" provided by crisis pregnancy centers is actually false and prejudicial. Scare tactics and shaming are two tactics employed. The fetus is always referred to as a person; the narrative with ultrasound films refers to the "boy" or "girl" in the uterus and interprets reflexive movement as if it were knowing and intentional. Horrible things are said to happen to women who undergo abortion, like severe depression, sterilization, and breast cancer. None of these is scientifically valid,[9] but they have the effect of scaring the uninformed and frightened woman who thinks these people are godly and would surely not intentionally mislead them.

Public Policy: The Critical Issue

The final solution sought by the anti-abortion forces is a legal ban or constitutional amendment that would prohibit abortion. Those who want to allow some exception for threat to the life of the woman are now being ostracized by the radical faithful. There seems no room for moderation. The most radical do not allow an exception for women's health since they argue that the fetus is "innocent," thus having higher moral value and deserving greater legal protection than the woman.

For anti-choice forces, abortion is the single-most important social and moral issue in America, as former President Ronald Reagan put it. In fact,

[8] Esther Kaplan, *With God on their Side* (New York: The New Press, 2005) 139.

[9] See Paul D. Simmons, "Post-Abortion Depression & the Ethics of Truth-Telling," *Christian Ethics Today*, Vol 9, No 3, Summer 2003, 21–24.

opposition to abortion serves as an umbrella for a cluster of social issues targeted by the coalition. The list includes denying civil rights to gays and lesbians, public funding for sectarian schools, encouraging Intelligent Design to be taught as science, promoting the submissive or at least inferior role of women in society and the right to proselytize in public schools and sporting events. Abortion remains the "wedge" issue, however, which drives the emotional fervency behind the social and political effort on the far right.

Abortion is finally the *only* issue on the agenda of this movement. If one is against abortion, and that absolutely so, one qualifies as a true Christian believer. Columnist Bettye Baye questioned Richard Land, executive director of the Ethics and Religious Liberty Commission of the Southern Baptist Convention, as to whether abortion is actually the top moral issue facing Americans. The backdrop was the devastation to New Orleans and the Gulf Coast in the aftermath of Hurricanes Katrina and Wilma. Land maintained the absolutist party line that abortion is the primary evil, while Baye thought both racism and poverty more important.[10]

Land's list is simply too short. It emphasizes abortion but pays little attention to major moral issues confronting millions of people in the world, including HIV/AIDS, racism, poverty, war, drugs, terrorism, health care, the environmental crisis, human population growth, and capital punishment. Others could be added. Wherever it belongs on the list, abortion is not number one, much less the only one. Neither virtue nor piety is measured by one's position on abortion.

My own take on the debate is that abortion serves as a smokescreen and diversion in order to get the attention of Americans off major social issues, including the use of power in foreign policy. The abortion wars are a type of social hysteria generated by radical ideology but serving to distract Americans from vital issues that need attention. Randall Balmer says abortion was suggested as a political ploy to unite disparate groups under a common, emotional banner. He says it was invented to take the focus off racism as a central moral concern in the elections of 2000 after Mr. Bush spoke at and implicitly endorsed the religious and social philosophy of Bob Jones University.[11] Actually, the fundamentalist war against abortion started much earlier, though it provided a perfect foil for the Bush campaign.

---

[10] Bettye Baye, "Ministers Differ on Whether Churches Should Target Abortion, or Poverty," *Courier-Journal*, 10 November 2005, editorial.

[11] Randall Balmer, *Thy Kingdom Come: An Evangelical's Lament* (New York: Basic Books, 2006) 16.

Abortion is also the code word for being moral while engaged in extremist actions—all other vices are covered and forgiven if one is sufficiently radical about prohibiting abortion. One might do anything else, but as long as one is against all abortions, one is moral and to be celebrated, honored, and promoted to a position of leadership.

All bets are off once a child is born, however. Horrible things can be done to women, the poor can be neglected and blamed for their poverty, violence can reign in the schools, millions can die of AIDS, and hundreds can be tortured in American concentration camps without one word of protest from those who claim to be the keepers of the nation's moral springs. It is a "bottom line" morality. Nothing counts but this one thing. A prostitute once remorsefully told her counselor, "I have done just about everything you can imagine," but then she brightened up and said triumphantly, "But I have never told a lie." So it is with the anti-abortion zealots: they have done many things that are wrong, but they have never approved abortion!

Candidates for office have been targeted for defeat or election on the single basis of their opposition to or support for anti-choice objectives. Some of the worst tyrants in history banned abortion, including Stalin, Hitler, and Caucesceau, which is a blunt reminder that being against abortion is hardly a reliable test of moral integrity or religious piety.

Anti-abortion goals are also portrayed as "pro-family," based upon the claim that they are protecting "innocent human beings" (i.e., the unborn, not people falsely charged with crimes). These groups are affluent, well organized, and aggressive and have shown considerable influence in determining the outcome of local and national elections. In spite of their minority status, they gained substantial power in Congress and various state legislatures, not to mention the federal government during the second Bush administration. As Esther Kaplan says, what was "once a marginal social movement is now driving politics in Washington."[12] Before the elections of 2006, a third of the House and Senate had voting records that were 100 percent consistent with the agenda of the religious right, and President Bush had appointed over 200 federal judges for whom the litmus test was a fervent commitment to overturn *Roe v. Wade*.

The cluster of issues under the abortion umbrella includes stem cell research and contraceptives regarded as abortifacients, that is, any chemical or barrier that prevents either fertilization or implantation. The morning-after pill (Plan B) and emergency contraception (EC) are thus also under

[12] Kaplan, preface.

attack. In every case, it is argued, the concern is for "the sanctity of human life," that is, its inviolability.[13]

The seriousness of the debate turns on the contention that the embryo (fetus, stem cell) is on par morally with the woman. By their logic, the "life" of the conceptus (from fertilization onward) is regarded as of "equal worth" with that of the woman or any other person. No distinction in moral value or legislative protection is made from conception to birth. When President Bush signed the bill banning so-called "partial-birth abortion," he called it the "partial delivery of a live boy or girl, and a sudden, violent end of that life."

The next step in the argument is to regard the destruction of an embryo or fetus at any point in pregnancy as murder. The biblical commandment against murder (Ex 20:13) is directly applied to abortion as if it were a divine command that settles the debate. Some even argue that abortion to save the woman's life is evil since "better two deaths than one murder."

In this line of logic, it follows that the moral law should be implemented by the civil and criminal laws of the country. For anti-choice groups, the law should protect the right to life and all property rights of the fetus as a citizen-person. (See Appendix V, "Respect for Human Life," Part I.1.)

A final cluster of arguments concerns what are regarded as the evil consequences that are sure to follow from laws permitting abortion, namely, sexual promiscuity and irresponsible sexual behavior, including non-marital sex and non-procreative sex within marriage. Albert Mohler, president of The Southern Baptist Theological Seminary, Louisville, Kentucky, and a vocal leader among anti-abortion fundamentalists, has called the Pill "the most immoral invention in history" because, he says, "it has given incredible license to everything from adultery and affairs to premarital sex, and, within marriage, to a separation of the sex act and procreation."[14] His statement sounds very much like the Vatican document "Respect for Human Life" (See also Appendix V, II.B.). Since when have Baptists in particular and Protestants in general opposed the use of contraceptives? But Mohler now wants them banned and pretends to speak for all "Christians" in America.

---

[13] See Pope John Paul II, *The Gospel of Life* (New York: Times Books, 1995) 95.

[14] Albert J. Mohler, quoted by *The Courier-Journal*, 30 May 2006, editorial. See also, John Paul II, *Gospel*, 25.

Religious Liberty: The Free Exercise of Power?

Mohler also argues that he embraces religious liberty. There is some truth in that claim since he emphasizes First Amendment guarantees for "the free exercise of religion." He skips over the establishment clause, which he wants to change, and goes straight to the actions that might be employed to ban abortion, deny gay rights, etc.

There is little doubt that the Christian right exploits the freedom to be religious and openly political. The conversion of America to their notion of righteousness is seen as a religious and social mandate. Three "Justice Sundays" were organized by evangelicals intent on banning abortion and securing legal prohibitions for such things as gay marriage.[15] The emphasis on Justice Sundays, however, had nothing to do with justice as portrayed in Scripture or, for that matter, in American jurisprudence. It had to do with "justices," that is, judges, who would preside at cases involving issues of importance to them at every level of court authority in America. Speakers specifically targeted the Supreme Court with the openings occasioned by the retirement of Justice Sandra Day O'Connor and the death of Chief Justice William Rehnquist. They wanted only rabidly anti-abortion judges to be appointed. The first Justice Sunday was designed to support Senator Frist's threat of "the nuclear option" should the Democrats filibuster against John Roberts, the President's nominee for chief justice.

Justice Sunday speakers made no mention of constitutional governance in America or what "equal protection" requires, but they spoke a lot about the Bible, as if it had a clear word on such subjects that could be normative for all Americans! The meanings of justice in Scripture were never explored for all their nuance and power. The sessions in Louisville were closed to the public, and only parishioners of two (mega) evangelical/ fundamentalist congregations were given tickets to attend. All attendees were admonished not to talk about the sessions with the press after the meeting. Speakers included Senate president Frist (who spoke via satellite from DC) and President Albert Mohler of Southern Seminary. There was little doubt they saw their cause as religious and felt that theocracy was the thing for America.[16]

---

[15] Justice Sunday I, was on April 24, 2005, at Highview Baptist Church, Louisville, KY. Justice Sunday II was in Nashville, TN at the Two Rivers Church, and Justice Sunday III was in Philadelphia, ironically called "Justice and Liberty for all." Further information is available on line.

[16] See *The Courier-Journal*, 1 May 2005, for the full text of President Mohler's speech.

Not all anti-choice activists are openly religious, thus trying to dodge the limitations of advocacy imposed by the establishment clause. Some argue, for instance, that their opposition to abortion is entirely secular. Bernard Nathanson, narrator of *The Silent Scream*, was once a Catholic priest and believed an embryo to be a person based on religious teaching. Now that he claims to be an atheist, however, he says his belief has nothing to do with religion.

Another is Don Marquis, who focuses not on the genetic beginnings but on the outcomes or teleology of human life. The great moral harm in abortion, he says, is that one is deprived of all the goods of life they otherwise would have experienced, that is, a future of value.[17]

Roman Catholic theorists typically appeal to natural law ethics, which seeks universal and absolute norms for guidance in questions of morality.[18] These principles or norms for conduct are thought morally binding since they are held to be the truth of God. No claim is made to special revelation, however, arguing that these laws are "natural" and thus available to all people, making them both universal and absolute. Presumably this establishes a basis for agreement on matters of morals and law among Christians, secularists, and people of other world religions. Roman Catholic theologians have most consistently developed this approach, though it can also be found in Anglicanism and Protestantism through the influence of Luther, Calvin, and more recently, Francis Schaeffer.[19]

There are three major problems with this approach to ethics and public policy. One is that there is a great deal of disagreement among natural law theorists, leaving us to wonder just what is either objective or absolute about this approach to "truth." Another problem is the arrogance and dogmatism that attends their conclusions. Once they discern the truth, it is to be accepted by everyone, no matter whether the arguments make sense or whether others agree or not. The claim is self-serving in the extreme since the theorist accepts no criticism or modification of the norm and then attempts to use the civil law to enforce it on others. (See Appendix V, pt. 3.) The third problem, explored in more detail below, is that safeguards of

---

[17] Don Marquis, "Abortion and the Beginning and End of Human Life," *Journal of Law, Medicine & Ethics*, Vol. 34:1, Spring, 2006, 23.

[18] Paul D. Simmons, "Theological Approaches to Sexuality: an Overview," in *Sexuality and Medicine Vol. II: Ethical Viewpoints in Transition*, Earl E. Shelp, ed. (Dordrecht, Holland: D. Reidel Publishing Co., 1987) 199–217.

[19] Francis A. Schaeffer, *The God Who is There* (Downer's Grove: IntervarsityPress, 1968).

religious liberty or the separation of church and state are ignored or attacked as oppositional to God's law.

Politicized fundamentalists claim to base their arguments on what the Bible commands. They then try to impose what they *say* the Bible teaches on all Americans. They argue that America is a Christian nation and that it should embrace the teachings of its founding document, the Bible. Their interpretations run counter to scholarly understanding, or seem only tenuously related to the biblical text, however. The claim that the Bible condemns abortion is impossible to document,[20] but it sounds good to the anti-choice constituency. Most Americans (see note 27 below) remain committed to the legal availability of abortion in spite of the heated rhetoric of a fervent but misguided anti-abortion movement. Political strategies and well-organized campaigns more than compensate for their lack of persuasive arguments. Former senator Sam Nunn (D-GA) captured the problem by saying that "they are seldom right but they are never in doubt."[21]

The Unthinkable: Banning Reproductive Choice

Those who support choice believe the woman, not the state, should make the decision about abortion or other reproductive matters. Republicans for Choice, for instance, is a conservative political group that believes the government should stay out of private sexual matters. Jerry Muller says a massive expansion of government bureaucracy would be required to manage personal choice. Further, it is the parents who are responsible for assisting when a teen daughter gets pregnant. Finally, he complains that right-to-life groups treat "life" as if it were the only good to pursue. He says the anti-abortion movement is anti-family, pro-big government, and has an unsupportable view of "life."[22]

There is extensive support for legal abortion among religious groups who emphasize beliefs about individual moral responsibility and direct accountability to God. The Religious Coalition for Reproductive Choice (RCRC) attempts to represent and consolidate the efforts of various religious groups (Protestant, Jewish, and Roman Catholic) "to safeguard the option of legal abortion." Planned Parenthood is a secular group that

---

[20] Michael Luo, "On Abortion: It's the Bible of Ambiguity," *The New York Times*, 13 November 2005, 1, 3. See also Paul Simmons, *Birth and Death: Bioethical Decision-Making* (Philadelphia: Westminster Press, 1983) ch. 2.

[21] Cited by S. Crabtree, "Senate loses its center in flood of retirements," *Insight on the News*, November 20, 1995.

[22] Jerry Z. Muller, "A Conservative Political Posture," (unpublished) 1.

supports legal abortion as an extension of the rights of reproductive choice and as a backup to contraceptive failure.[23] Those who support abortion rights do not advocate or encourage women to obtain abortions, but argue that a woman should have the legal option when it seems best to terminate a pregnancy[24] They work on two fronts: one to keep abortion legal, the other to reduce the causes that contribute to problem pregnancies. One or more of the following arguments is usually made by the supporters of choice.

First, the woman, not the fetus, is the person at stake and whose interests must be preserved.[25] Tragically, the woman is discounted as a moral agent, while the fetus is elevated in moral value by anti-choice arguments. At best, however, the fetus is *potentially*, but not *in fact*, a person. This is especially true in the earliest stages of gestation. Simple stages of cell division and organ differentiation hardly make one a person. Potentiality is not actuality, just as acorns are not oak trees. Personhood can be meaningfully applied to a fetus only with viability, or the ability to live outside the womb. Only after this stage of biological and neurological maturation might a fetus be regarded as a "person," on commonsense or legal grounds.

Second, the rights of women are denied by laws that narrowly limit the availability of abortion. The more attention is focused on the fetus, the more the woman becomes invisible—of no importance in the calculus of moral or human rights. Health and Human Services secretary Tommy Thompson absurdly extended health services to "fetuses and embryos," without any mention of benefits to the mothers. The woman was lost in his rhetorical flourish. There is ample evidence of a war against women in the second Bush administration. Women are denied medical care when they cannot obtain an abortion legally but are forced in desperate circumstances to seek help from illegal abortionists. Worldwide, women have suffered from laws similar to the "gag rule" in the Mexico City policy that deprives women of any information or conversation about abortion.[26] Planned Parenthood programs, interventions, and educational approaches have been de-funded both nationally and internationally as a matter of policy. The anti-woman

---

[23] See the history of the fight for reproductive rights at www.plannedparenthood.org.

[24] See Beverly W. Harrison, *Our Right to Choose* (Boston: Beacon Press, 1983).

[25] Bonnie Steinbock, "The Morality of Killing Human Embryos," *Journal of Law, Medicine & Ethics*," Vol. 34:1, Spring, 2006, 26–34.

[26] See P. G. Stubblefield and D. A. Grimes, "Septic Abortion," *NEJ M*, 1994 (331) 310–314, who document the worldwide problems of women who die from botched abortions and bad information.

bias extends the power struggle for women against men who pass restrictive laws but have never known the threat and burden of an unwanted pregnancy. The net result of anti-choice law is to force the woman to carry a pregnancy to term against her will.

Third, supporters of choice argue that the moral issue in abortion should be separated from legislative control. The ethical issue is a matter of personal judgment based on religious teachings and moral values. Those who believe abortion is wrong should not be required to abort for any reason; those who believe it is acceptable should not be prevented from obtaining an abortion (prior to viability). Further, the complexity of particular cases (genetic deformity, rape, incest, emotional state of the woman, etc.) is so great that a simplistic law cannot deal with all exceptions, or the law becomes so complex and cumbersome that the woman is subjected to undue harassment in checking to see whether her case "qualifies." Better to leave the decision to the woman who, after discussions with her physician and other people of importance to her, can weigh the issues and decide. Apparently, a vast majority of Americans support this argument.[27] Those who are trying to outlaw abortion are trying to impose a minority view on everyone.

A final argument relates to the issue of religious liberty or freedom of conscience.[28] Most church-state groups have sided with the court's decisions to legalize abortion, not because they or their constituents believe abortion is always right but because they believe the First Amendment is abridged when religious opinions or metaphysical abstractions are implemented as law.

The definition of "person" in anti-choice arguments, for instance, is a matter of metaphysics or theological teaching. The only basis for believing that an embryo has the worth of a person is to posit some notion of an unseen metaphysical reality not provable by science. The contention is

---

27. In 1998, The Alan Gutmacher Institute, which tracks statistics on social attitudes and practices regarding abortion reported very little shift in attitudes toward abortion since 1973. In 1975, 21% thought abortion should be legal, and 22% held it should be illegal under any circumstances; 54% thought it should be legal under certain circumstances. In 1997, 26% held it should be legal under any circumstances, 17% illegal under all circumstances and 55% thought there should be some restrictions but legal. *The Houston Chronicle*, Mon. Jan 19, 98, 1A, 8A.

28. P. S. Wenz, *Abortion Rights as Religious Freedom* (Philadelphia: Temple University Press, 1992).

usually that the embryo has a soul,[29] which is fine as a religious belief but problematic for the First Amendment. Laws in America are not to be based upon sectarian or religious views by which everyone must live. America is a pluralistic society, and religious groups are deeply divided on the abortion question. Why, then, should the government intervene to criminalize the behavior of those who base their behavior or choices on profoundly held religious commitments? The Constitution guarantees freedom *from* religious dogma as well as freedom *for* religious belief. Anti-abortion laws pose significant issues for governmental "establishment of religion."[30]

As Rabbi Shira Stern puts it, "The debate surrounding reproductive choice speaks to one of the basic foundations upon which our country was established—the freedom of religion. It speaks to the right of individuals to be respected as moral decision makers, making choices based on their religious beliefs and traditions as well as their consciences."[31]

History, Law, and Public Policy

Profound religious beliefs about the moral acceptability of abortion thus continue to generate heated debate in political, medical, and religious circles. At stake are oppositional perspectives that lead to very different moral conclusions regarding the termination of an unwanted pregnancy and the role of law in regulating these choices. Supreme Court decisions have dealt with and helped to focus the basic questions about abortion.

A foundational decision for the current debate is that of *Griswold v. Connecticut* (1965), which overturned anti-contraception laws based largely on the crusade led by Anthony Comstock. The Supreme Court ruled that there is a right to privacy that includes choices regarding the use of contraceptives. Prior to *Griswold* most states outlawed the sale or distribution of contraceptives. Legislation against contraceptives had strong backing from the Catholic Church.

Roe v. Wade followed in 1973, ruling that the right to privacy was sufficiently broad to encompass decisions about abortion. The court refused to rule on "when life begins," saying that it was in no position to decide the question when philosophers and theologians are so deeply divided on the

---

[29] F. Fukuyama, *Our Post-Human Future: Consequences of the Biotechnology Revolution* (New York: Farrar, Straus and Giroux, 2002) 161.

[30] See Paul D. Simmons, "Religious Liberty and the Abortion Debate," in *Journal of Church and State*, Vol. 32 (Summer 1990) 567–584.

[31] Rabbi Shira Stern, Letter supporting President Clinton's veto of HR 1122, September 10, 1998.

issue. *Roe* regarded abortion as a matter between a woman and her physician. Limits on abortion were related to the stage of gestation at which the decision was considered. In the first trimester, the woman's decision is her own; during the second trimester, abortion should be regulated in a way that protected the woman's health; and in the third trimester (following viability), the decision could be limited to those cases in which there is serious threat to the woman.

In *Webster v. Reproductive Health Services* (1989) the court ruled that limitations on abortion could be established by the states. Missouri forbade abortion by any physician who received public funds and prohibited public funding for abortion services. It also provided incentives for childbirth and required the woman to be informed of fetal development—the hope being that it would deter her decision.

Planned Parenthood of Southeastern Pennsylvania v. Casey (1992) affirmed both *Roe*'s legalization of abortion and *Webster*'s limits on the practice. The court rejected arguments that a fetus was a person while strongly affirming that it is the woman who is constitutionally protected. It declared that no state can ban abortion. Four states that had already banned the procedure were thus rebuffed by the court.

Public policy now follows the patterns outlined in this series of Supreme Court decisions. In effect, the court has recognized the abortion debate as a matter of religious and philosophical opinions that are strongly held but not subject to legal resolution. Issues in religious liberty were not directly addressed, but seemed implicit, in the discussion and the concerns for the woman as person. The net effect of the various decisions is that each woman is supposedly free to believe and act according to her own deeply held religious beliefs and moral values. Further, there were definite limits set around the ability of organized religious groups to establish a narrow theological view of fetal personhood as public policy.[32]

Even so, the court left considerable latitude for states to impose hindrances or limitations on the availability of abortion. The anti-abortion crusade has thus turned to state legislators and a cooperative, committed US President who pursues the cause by executive orders and federal appointments. It has been a clever but harmful strategy in the name of a moral crusade the foundations of which are terribly problematic, if not an overt effort to rewrite the Constitution.

---

[32] See Paul D. Simmons, "Religious Liberty and Abortion Policy: *Casey* as 'Catch–22'," *Journal of Church and State*, Vol. 42 (Winter, 2000) 69–88; also see my *Freedom of Conscience* (Amherst, NY: Prometheus, 2000) 240–262.

The Sacrifice of Becky Bell

The tragic story of Becky Bell illustrates the problem of compromising liberties in the name of religious metaphysics. Bell was a sixteen-year-old Indianapolis woman who was reared in a devout Roman Catholic home. Her mother and father were strongly against abortion and insisted she be sexually abstinent until marriage. There are good reasons for chastity prior to marriage, but passion won out over her best resolutions, and Becky became pregnant. She had thought that using contraceptives would mean she was bad, both in her own eyes and in the eyes of the church. At other times, she saw herself as a good girl and thought God would protect her even if she did not use protection.

When a test showed she was pregnant, Becky talked with her boyfriend. They decided they were not ready for marriage and a child, but she also felt confused and desperate. She felt too much shame and fear to face her parents. She tried to find a judge to implement the "judicial bypass option" under Indiana law, but could not. In the years since the law had been passed, not one judicial bypass had been granted in Indianapolis. So Becky found an abortionist. She did not know who it was, where it took place, or what credentials or medical experience the abortionist had.

Becky wound up with sepsis. A terrible infection raged in her body. She developed a high fever, and death followed after three days. Her parents were heartbroken. They wished they could have those conversations about abortion and sex with Becky all over again. A major part of their grief is that Becky did not understand that she could tell her parents anything and that they would support her regardless of what had happened.

Now her parents have launched a campaign to have states make laws that are more friendly toward teenage women. The Bells see the judicial bypass option as unrealistic, unworkable, and inhumane. They believe a structure needs to be in place where young women can feel safe and protected.[33]

---

[33] Notes from an address by Mrs. Bell to the Kentucky chapter of Religious Coalition for Reproductive Choice, Louisville, KY.

Guidelines for Public Policy

There is likely little possibility that the opposing sides in the abortion debate will reach détente in the near future. Even so, there are at least four points to be made that should provide grounds for civil discussion, if not common agreement.

First, the issue for public policy is a definition of personhood consistent with scientific views of gestational development and appropriate in and for a pluralistic society. Religious commitments motivate both sides in the debate about the moral acceptability of abortion and whether the practice should be legal. A variety of perspectives can be found as to when a developing fetus should be regarded as a person, which reflects the religious pluralism of America. No religious group has a position that is uniformly and unanimously accepted even by all members of its own group. Women who have abortions belong to a variety of religious traditions, including those that have most consistently and dogmatically condemned abortion. Women must make their way in terms of their own religious beliefs; they cannot be categorized in terms of the religious organization to which they belong.

Second, there is and should be a right to privacy. The *Griswold* decision did not simply invent such a right, nor did it need to find "a right to privacy" in so many words in the Constitution. The court summarized and succinctly stated a right expressing what most Americans take for granted: that there is a zone of thought, beliefs, and ideas that is uniquely personal and insulated from invasions or violations by government or others. All groups in the debate appeal to the power of their own inner convictions. Privacy points to and preserves the interiority and profundity of one's moral values and religious beliefs; it is not just what people do behind closed doors. The courts have consistently showed deference to religious motivations in actions and provided protections for individual conscience, which is synonymous with religious convictions. Ethically, recognizing and respecting privacy is another way of speaking of regard for the person.

Privacy rights also give meaning to the notion of freedom *from* religion. Those who want no religion at all should be respected and protected. Aggressive religions should not be permitted to coerce others in matters of conscience or religious belief.

Third, public policy governing abortion should serve the interests of women and families in all religious groups. The question at law is whether the definition of personhood is reasonable or logically problematic, and whether it enhances or restricts protected personal liberties. The first principle of religious liberty is that laws will not be based upon abstract metaphysical speculation, but will be fashioned through democratic processes in which every perspective is subject to critical analysis. Appeals to "God says" or to "natural law reason" expose a claim to special knowledge not everyone shares or can agree upon. Such arguments are either linguistic chicanery or an intellectual dodge, typically resorting to structures of authority when persuasion fails. No sectarian religious answer should be imposed on all Americans.

Adopting a notion of the embryo that would criminalize abortion would be an exercise in the establishment of religion and result in widespread denial of women's rights. Government has no business settling the thorny questions and debates in religion and philosophy. In spite of claims to the contrary, all notions of an objective truth regarding the status of an embryo are without foundation. The embryo is highly valued for a variety of reasons, or it is feared and loathed for reasons that are particular to the woman under adverse circumstances.

Rabbi Shira Stern is to be protected and respected as well as the most ardent anti-choice woman. Each approaches the issue because of ground-of-meaning or religious beliefs that provide warrants and justifications, or prohibitions and restrictions for her actions. *Roe* allowed each woman to be faithful to her own religious beliefs and commitments. The woman whose conscience is formed against the procedure is in no way coerced or required to have an abortion, no matter what the circumstances of her pregnancy. She can be true to her own conscientious convictions.

The same should be true for those like Shira Stern who reflect and act upon the religious tradition from which they come and the particular circumstances of her pregnancy. Rabbi Stern's religious leaders and community supported her decision to abort. It would have been a severe miscarriage of justice and an imposition of an alien and odious religious opinion to have forbidden her action.

Chapter 11

MEDICINE AND THE DEMONIC: SHOULD PSYCHIATRISTS ENGAGE IN EXORCISMS?

> Then he...gave them authority to drive out unclean spirits and to cure every kind of illness and infirmity.
>
> —Matthew 10:1

> Jesus gave the responsibility for healing and casting out demons...to every Christian."
>
> —Paul Tillich, *The Eternal Now*[1]

Few issues test the relation of faith to science more directly than the practice of exorcism, or the effort to cast out demons. Exorcism is premised on religious beliefs about devils and demons, but the aim is to relieve symptoms that pertain to health. More accurately, the symptoms are those of disease or mental disorders that are confounding the efforts of traditional medicine or psychiatry. When frustration turns to loss of hope in healing by medication, people who care for the afflicted person may turn to paranormal explanations. For centuries, the idea of demonic possession has been suggested by those who believe strongly in demons and that illness may best be explained and treated as a matter of possession. Medicine claims and seeks a scientific explanation; exorcism claims and seeks a cure based on religious ideation.

The idea of demonic possession seems a current fascination. A check with a single Internet site yielded no less than seventy-five references dealing with Satan and/or his minions. The site did not include books by authors who do not accept the existence of demons, however, like the late

[1] Paul Tillich, *The Eternal Now*, 59.

Carl Sagan.[2] My own interest in the topic has had several dimensions. One is my work at the medical school, where the question of demon possession and psychosis is sometimes posed as both an academic and as a clinical question. Scott Peck, the late psychiatrist and popular writer, provoked pointed questions for me with his chapter on the subject and his participation in exorcisms. This chapter intends to do precisely what Peck asked, namely, to take seriously the issues involved in exorcism and the assumptions that drive current interest in the subject.

Two things seem important when dealing with the subject of exorcism. First, the phrase "casting out of demons" (Matt 10:8) may be a meaningful metaphor for contemporary medicine and religion. Second, a distinction needs to be made between figurative language and entities to which we might attribute ontological status. There is an all-important distinction between a literary device and an ontological reality. An examination of the question of Satan and demons in Scripture, history, and science may provide guidance that will enable thoughtful people to work through the issues in a way that involves the insights of both science and religion. Exorcisms like that described by Peck pose substantive problems for professional ethics. A partnership between medical science and religion is needed to help clarify the nature of evil and the problem of human suffering. Spiritual realities are clearly involved in human illness, and medicine cannot do its healing work without recognizing such dimensions in pain and suffering.

A Story of Demonic Possession

Scott Peck's story of an exorcism and Paul Tillich's challenge to employ the metaphor of "casting out demons" in contemporary ministry help to frame my approach to the issue. Peck was a popular psychiatrist and inspirational writer. Tillich was an eminent theologian whose aim was to integrate religious perspectives with scientific insights. Peck's book *People of the Lie*[3] includes a chapter, "On Possession and Exorcism," in which he accepts and embraces a world inhabited by demons. Tillich had also spoken of "exorcism" and the "casting out of demons" to a graduating class at Harvard Divinity.[4] Peck and Tillich are using similar language and share certain profound concerns for the health and well-being of persons in the world, but

---

[2] Carl Sagan, *The Demon-Haunted World: Science as a Candle in the Dark* (NY: Random House, 1996).

[3] M. Scott Peck, *People of the Lie: the Hope for Healing Human Evil* (New York: Simon and Schuster, 1985).

[4] Paul Tillich, *The Eternal Now* (New York: Charles Scribner's Sons, 1963) 58.

did Tillich agree with Peck about the existence of demons (or Satan)? Was he encouraging psychiatrists to lay aside the art and science of medicine and use a more hands-on approach to the evil confronted in a young woman who was mentally ill? Hardly. Tillich's approach was to demythologize the notion of demons. Even so, he retained the power of metaphorical language while rejecting any metaphysical reality of demons, which he regarded as a figment of a fertile, fearful, and fragmented imagination. He never ceased reminding the religious of the subtle, but powerful, temptation to confuse superstition with faith.

Peck and Tillich are thus worlds apart. Peck dealt with demonic possession as a legitimate explanation for certain human diseases, particularly in the form of mental illness. There is such a thing as "genuine possession," he claimed.[5] His interest in the subject led him to "go out and look for a case." The first two were "disappointments," he said, but the third "turned out to be the real thing." He attended these "successful exorcisms" and insisted they were not cases involving "minor demons," but actually "cases of *Satanic* possession." He said "I now know Satan is real";[6] he had met "Satan face-to-face."[7] He had confronted the head honcho, not his little league trainees.

Peck refused to describe the cases or give an in-depth presentation on the subject because of what he said were "a number of compelling reasons." Each case, he said, was "extraordinarily complex." He did indicate that the source of his knowledge of satanic possession was Malachi Martin's book *Hostage to the Devil.* He became convinced that counselors are "seeing such cases, whether they know it or not."[8] He argued that the two exorcisms in which he engaged were psychotherapeutic processes, both in method and in outcome.

He also admitted that engaging in an exorcism involves some problematic uses of force and power. Exorcism "is a dangerous procedure,"[9] which he compared to "radical surgery."[10] The means are acceptable, he says, only because they are "legitimately, lovingly available in the battle

---

[5] Ibid, 183.
[6] Ibid.
[7] Ibid., 184.
[8] Ibid.
[9] Ibid, 185.
[10] Ibid. 187.

against the patient's sickness."[11] The purpose is to uncover and isolate the demon within the patient so that it can then be expelled.

He also says the patient "must sign an elaborate authorization form." They must "know exactly what they are letting themselves in for"! Peck does not deal with the nature of informed consent for the mentally ill, nor does he explain how people who are overwhelmed by demonic possession could possibly understand either their condition or anticipate what "they are letting themselves in for." Both would be necessary conditions for a valid patient consent. Peck admitted that a guardian (in this case, presumably, the parents) might have to make a proxy decision for the patient. Calling such attention to details both "idealistic and impractical," however, he admitted that he would forgo even the guardian permission "in desperate instances."[12]

To safeguard patient safety and well-being, Peck said, accurate and complete record-keeping were required, though readers are not given critical details. Documentation is necessary, but so is "love,"[13] we are told. Love enables the practitioner to "discern between interventions that are 'fair' and necessary and those that are manipulative or truly violative." Peck called exorcism a type of "psychotherapy by massive assault," admitting the actions were heavy-handed but authorized by the *Rituale Romanum*. Furthermore, "the exorcism might fail and (the patient) might even die as a result of the procedure."[14] There is even danger to the exorcist and other team members, he says.

Peck says possession is "in addition to" the patient's mental illness. Both patients he interviewed showed what he admitted were "multiple manifestations of routine mental illness such as depression or hysteria or loosening of associations."[15] He insisted, however, that the work of Satan differed in each case. In one, "the secondary personality *desired* to confuse (the personalities)."[16] In the other, even though the deliverance was unsuccessful, "an utterly evil persona temporarily emerged," only to be followed by a severe and life-threatening illness during which the "voice of Lucifer" was heard. Peck's conclusion was that this experience with exorcism

---

[11] Ibid., 185.

[12] Ibid., 187.

[13] Ibid., 188.

[14] Ibid., 189. Seven deaths in recent years have been related to exorcisms, including that of a 23 year-old nun, who was found dead in her convent room. Fr. Daniel P. Corogeanu led the exorcism but wound up on trial for murder. See en.wikipedia.org/wiki.

[15] Ibid., 192.

[16] Ibid., 193.

served "to strengthen our suspicion that the demonic was playing a major role in the person's illness."[17]

Peck says he saw an expression on the patient's face that could be described only as satanic. It was, he said, "an incredibly contemptuous grin of utter hostile malevolence." He was so intrigued that he "spent many hours in front of a mirror trying to imitate it without the slightest success."[18]

In the second patient, Peck claims Satan appeared as a writhing snake of great strength, viciously attempting to bite the team members. The eyes were hooded with lazy reptilian torpor, except when the reptile darted out in attack, at which moment the eyes would open wide with blazing hatred. This "serpentine being" gave Peck "a sense of a fifty-million year old heaviness."[19] These "mystical" events, he declared, are not just isolated, but somehow almost cosmic happenings.[20]

Finally, Peck relates his discussion of demonic possession to clients he has discussed elsewhere in the book—Bobby's and Roger's parents, and Sarah and Charlene. He wonders whether they are not "thoroughly evil people [and] cases of perfect possession?"[21]

Through all this material, both descriptive and analytical, Peck calls himself a "hardheaded scientist"[22] and ends by proposing a "centralized data bank and study center" that could serve to teach with the use of videotapes of exorcisms.

Speaking in Parables: A Brief History

Peck's foray into demonic possession has touched a nerve in the contemporary psyche. The topic seems to be one of near-universal interest. At a minimum, we can agree with Peck that the subject of demons and possession appeals to fears that "do lie too deep for tears." Who of us did not watch the televised exorcism performed on a fifteen-year-old girl by ABC News?[23] It was treated with all the dignity and acceptance usually

[17]Ibid., 194.

[18]Ibid., 196.

[19]Ibid.

[20]Ibid., 200.

[21] Ibid., 210.

[22] Ibid., 195.

[23]ABC news, "20/20," hosted by Hugh Downs and Barbara Walters, April 5, 1991. The report on Exorcism was by Tom Jerrold about a young woman whose symptoms included recurrent psychotic episodes. Hugh Downs called it "a mesmerizing session," while Barbara Walters wondered why the church would allow such an episode to be filmed.

accorded items worthy of a television audience! The exorcists went through all the rituals prescribed by the ancient *Rituale Romanum*. And as Peck claimed in his case, there did appear to be a type of catharsis, release, or breakthrough on the part of the patient, but Peck admits that it was impossible to tell whether the temporary relaxed state was from exhaustion or from the exorcism. Follow-up studies indicate the girl he treated is still struggling with chronic psychosis.

Peck's descriptive tale of people possessed by Satan tends to capture the imagination. It may even appeal to the mind that wants to believe or finds an explanation for the mental illness that has deprived a loved one of rational processes or healthy emotions. The reader might even identify with responses when watching *The Exorcist* or any of its sequels; whose sleep patterns went totally undisturbed afterward? The mind is a picture gallery, not a debating hall, and strong images have a way of entering our patterns of thought unbidden or during sleep, or when made anxious by the dark. The question begins to haunt us as to whether there might be something to all this about demon possession? After all, such stories seem terribly convincing.

Furthermore, the world's great religions often engender and enforce human fears of the unknown and mysterious with references to evil spirits, beings, or forces. In the Mesopotamian creation stories, Leviathan, the great dragon, is portrayed as the spirit of chaos and darkness. The Akkadian and Sumerian cuneiform texts have invocations against the malice of demons. Iranian (Persia) dualism (eighth century BC), which influenced so much Jewish thought during the Graeco-Roman Period, has its Angra Mainyu (Ahriman), the foe of the supreme god Ahura-mazda (Ormazd), or of Druj, the spirit of deceit. Those two supreme gods—one good, one evil—are locked in an eternal battle in the religious drama portrayed in the sacred text.

Satan and the Scripture

Neither the Hebrew nor Christian Scriptures have a full-blown satanology. Even so, there is a "Satan" figure that appears in certain biblical stories. The serpent in Genesis is sometimes thought to be an evil being with supernatural powers, but in the biblical story, the serpent is simply a clever creature able to tempt Eve and Adam. It is hardly the Satan of contemporary interest based on medieval religious literature with its manuals about demon possession. The snake is "the Tempter" and is more nearly a symbol of the evil of which humanity is capable than anything like a totally evil being. The serpent of the garden is a creation of God, as are all other creatures of earth. The Tempter simply suggests—or plants the idea—that death might not be the immediate result of eating the forbidden fruit (cf. Gen 3:4ff.). The text employs a literary device in an effort to express and portray the nature and origin of the subtle suggestion—of temptation—or the idea that what is thought to be true may be, in fact, an error.

The "Satan" in the book of Job is the "Adversary," the one against people. He is part of the divine entourage—the messenger who checks out those who claim to be just or righteous. He is in a joint venture with God. A similar being appears in Zechariah 3:1–2, who challenges the fitness of Joshua ben Jozadak to act as a high priest at the time of the restoration from the Babylonian exile. Again, he is "the Satan," or "the accuser," but not an evil being.[24]

The belief in a being or power as the source of evil certainly seems to have its origins in the Old Testament. Belief in Satan was an outgrowth of theodicy, the question of how to reconcile the reality of evil with the righteousness of God. The notion of an evil power became a logical corollary to the revelation of the nature of God. If God is good, merciful, and loving—a God of righteousness and mercy (*chesed*)—then how do we explain the experience of evil in nature and history?

The image of the serpent in the garden and the adversary in Job removes the problem one step from God but ultimately still leaves God responsible for evil. The first step in Hebrew theodicy is found in the idea that "an evil spirit" tormented Saul (1 Sam 16:14–16) and then was said to have entered Saul when he threw a javelin at young David (1 Sam 16:23). Later, David was said to have been incited to the sin of taking a census by an evil spirit (1 Chron 21:1). "Evil spirit" is a common noun, which denotes a spirit or attitude, a type of personification of a human frailty.[25]

The New Testament speaks of Satan thirty-three times, but there is no clear teaching either as to the meanings of the term or of the origins of a Devil. Satan simply appears in the writings. John says he is the source of evil: "a liar and the Father of all murderers" (John 8:44). The point is theological and ethical: lying and killing do not have their origins in God, who is the source of truth and life. In short, the Devil and the demonic seem to be ethical concepts, not metaphysical beings. Each of the works of Satan listed in the New Testament is a moral concept.

---

[24] See Elaine Pagels, *The Origin of Satan* (New York: Random House, 1995) 39.

[25] An earlier account of the same incident says that it was Yahweh who incited the king (See 2 Samuel 23:1).

Those concepts were personified as a way to understand and relate redemptively to their influences and consequences in human life. The more personal and intractable suffering is, the more personal the antagonist appears, but the literary device of personification should not be confused as an identification of a person or a supernatural being.[26] Naming is a way of controlling; until an entity is named, it has power over the individual or group. Once named, it can be brought under the power of the one who named it.

Satanology in the Middle Ages

The development of a full-blown demonology awaited the Middle Ages. Demons, as the legions of Satan, are the products of the imagination; the clever inventions of people who had the time, inclination, and metaphysical mindset to catalogue every vice or virtue known to human beings.[27] John Milton probably did more to enliven the Western imagination with Satan and his legions than all Scripture taken together.

For every virtue, an angel, for every vice, a demon was named—those little devils or "lesser spirits" who serve in Satan's kingdom, that is, the world responsive to Satan, the source of all evil. The *incubi* and *succubi* were thought to be libidinous spirits, for instance, who seduce virtuous women and men in their sleep. The lord of all demons is *Satan.* He reigns supreme. He is both like and unlike the king of the kingdoms or the prince of the region. He has power; he rules; he has servants and imps who do his bidding. He is the one against people—testing them, tempting them, doing evil things to them. Possession was the manner in which Satan's imps did the bidding of the chief malefactor.

Witches were another insidious form in which "possession" supposedly took place. These were people—men and women, warlocks and witches—who were thought to be in league with Satan and thus able to cast spells and curses on others. John Bunyan, an early champion of religious liberty, once testified against a Quaker who was being tried as a witch. Bunyan supported the charge that she had turned a man in Cambridge into a cow.

The publication of *Malleus Maleficarum*, or "Witches Hammer," by Pope Innocent VIII in 1484, gave official recognition of demonology by the

---

[26] Pagels, *The Origin of Satan*, says the term was used to characterize one's enemies as under the control of evil transcendent forces.

[27] See Robert Masello, *Fallen Angels...and Spirits of the Dark.* (NY: Berkley Publishing, 1994) for a catalogue with descriptions of their duties.

church.[28] It is a handbook that describes the methods for identifying, examining, and torturing witches. Many of those identified were people with mental or behavioral disturbances, such as those in Peck's stories, of course. Others were dissenters or unorthodox believers. Their non-compliance to church orthodoxy was attributed to the influence of Satan. For that, they were condemned to death while church officials quoted Exodus 22:18: "You shall not allow a witch to live."

So much for psychotherapy and healing—or toleration of dissent from orthodoxy! *Malleus Maleficarum* represents the highly negative influence of the dualistic and spiritualistic model of mental illness. That notion still persists as a way to understand and treat the mentally deranged. A critical study of the history of demonology should provide an important corrective to wild imagination and misguided superstition.

Scripture and the Sensational

Some insist that the New Testament provides a clear and unavoidable doctrine of Satan and his legions of demons, but a careful study of Scripture hardly supports such a contention. At a minimum, biblical perspectives are more nuanced than we are often led to believe. The exorcism in Mark 5, for instance, is significantly different than those described by Peck. In Mark's story, Jesus "spoke sternly" to the demon[29] to come out of the man. Jesus hardly engaged in some elaborate ritual that might possibly injure or kill the man or those around him.

Further, any approach to demonology must make sense of Jesus' statement to Simon Peter: "Get thee behind me, Satan" (Matt 16:23; Mark 8:33; Luke 4:8). Jesus is actually calling Simon "Satan"! The occasion is recorded by all three Synoptic writers. It makes a point that is basic and indispensable to understanding the nature of the demonic and its power in human life. Notice the following: (1) Simon was obviously not Satan, even though Jesus calls him *satana*; (2) Simon embodied and articulated an idea or belief that was not "from God." That is, it represented a contradiction to God's will for Jesus. To follow Simon's suggestion would be for Jesus to

---

[28]Kurt Koch, *Demonology Past and Present* (Grand Rapids: Kregel Publications, 1972) 31. Koch calls this document "infamous" and condemns it as "responsible for the deaths of thousands of innocent people...."

[29]The story indicates that, when asked its name, the demon replied "We are legion," that is, six thousand. The symbolism of the number has enormous significance. Legion is not a name as such, but apparently a way of alluding to the overwhelming nature of the man's illness.

follow "Satan"; and (3) Simon represented a temptation to Jesus (much like those in the wilderness) to turn back from pursuing his understanding of what God willed for him to do.

In that sense, Simon could have been said to be "possessed" by Satan, but only in that sense. He certainly did not display the features of possession outlined in the *Rituale Romanum*, or those of witches in the *Malleus*. This incident underscores a truth that undermines all arguments for possession by demons or by Satan.

Evil and Human Free Will

People who do evil are embodiments of "satanic" power, but they are hardly "possessed" by Satan or demons. No approach to evil can be accepted that in any way diminishes or dismisses personal responsibility for evil actions. The problem with "possession" theology is that it falsifies the realities of the human situation regarding moral freedom and personal responsibility for evil actions. It tempts people to look to superficial answers for the enigmas of disease and disfigurement, of madness and derangement, and it gives a convenient excuse for those who try to diminish human moral evil. It is a Flip Wilson theology that "the devil made me do it."

No amount of speculation about demon possession can dismiss the moral horror or human responsibilities of a John Gacy, Jeffrey Dahmer, or Charles Manson and other serial killers who are certainly "possessed of evil," but they are not possessed by demons or Satan. The same can be said for notorious tyrants like Stalin, Hitler, Pol Pot, and Idi Amin. (The list could go on and on in political circles.) They were also "possessed" of evil in the sense that they embodied a spirit or attitude that contradicted the will of God for human life.

Ethically, the issue is the human capacity for doing evil. It is not a metaphysical being or ontological reality, but the nature of the human situation with which people struggle. Institutional religion and religious leaders are also tempted and subject to becoming embodiments of evil. Being pious or religious does not exempt one from the power of the demonic, but may actually increase the propensity toward the misuse of power. We need only think of the Inquisition, the Crusades, and the witchcraft trials.

Simon Peter was part of the inner circle of the followers of Jesus, but succumbed to the temptation to try shortcuts in pursuing the kingdom. Even the appearance of the Antichrist in Scripture is not one who is a secular threat to God's kingdom. The Antichrist is defined in religious terms, and even claims to be a "minister" of God!

Religious zealotism takes on the visage of the demonic. The witchcraft frenzy that swept Britain for over a century is a case in point. Thousands of good people were condemned to death by church leaders. One of the "witch finders" was a man named Patterson, whose flaming red hair added to the terror he was able to incite among the people. The test for whether one was actually a witch was simple—he or she would have no sensation of pain. Patterson used a probe with a retractable needle. When testing those he wanted to find guilty, he simply retracted the pin. No pin, no pain. The fact that they felt no pain was used as evidence of their sorcery. Needless to say, he was able to intimidate and terrorize simple people for years until he was exposed for treachery.

Another notorious example of religion succumbing to evil is the Inquisition, which lasted for nearly 600 years. It was led in part by the Franciscans, who were fierce defenders of the orthodox faith. So successful were the Inquisitors that the mere accusation of heresy was tantamount to being found guilty. Properties were confiscated, and "the guilty" were imprisoned, drowned, burned, or hanged. Through such a demonic process, the church accumulated enormous wealth and destroyed enemies as well as potential reformers.[30]

The demonic is also apparent in cultic fanaticism. We need only think of the massacres associated with Jim Jones in Guyana and the David Koresh catastrophe near Waco. The late William Temple had it right in saying that the first object of prophetic religion is the criticism of bad religion. The first task of the church is to be a critic of its own processes, perspectives, and misuse of power. The church should know of the demonic and be able to avoid it, not become subject to it. The ultimate evil is a religious evil. The Antichrist is demonic religion that perversely represents the truth of God but deceives the innocent to their own destruction.

Questions for Peck

With this background, certain pointed questions should be raised of Scott Peck regarding his decision to engage in an exorcism. To begin with, note that Peck's account of demonism and possession posed more questions than

---

[30] See Edward Burman, *The Inquisition: Hammer of Heresy* (New York: Dorset Press, 1992) who notes the linkage made by the church between heresy and sorcery (89) and the enormous financial benefits for the Church and Inquisitors from the trials of wealthy "heretics" (105).

it resolved. We are left with many burning questions: How did he know he was dealing with demons? How does "satanic" differ from "demonic" possession? How does one decide a person is "possessed" instead of mentally ill? Is this legitimate psychiatry, or does it border on, if not constitute, malpractice?

While reading *The Lie*, I could hardly believe I was reading a book by one of the best-known and most successful psychiatrists of our time. Peck gained a considerable following by writing in the areas of popular psychology, ethics, and healing, but his chapter on exorcism raised enormous questions for medical ethics pertaining to appropriate and acceptable interventions by professionals for the mentally ill.

The question was almost unavoidable as to whether someone other than the patient had been "possessed" by "the demonic"! Peck proposed that exorcism be taken seriously as a treatment of psychotherapy! It may also have been sensationalism parading as therapy for the high-stakes game of being number one on the bestsellers' list. At a minimum, it is incredible that a man of such stature would delve into a subject so wrought with superstition and discredited by modern science.

A number of Peck's statements posed questions requiring scholarly attention. For instance, where are such cases of possession "well described in Christian literature under the name of Satan"?[31] He admitted that his knowledge of historical theology was at best superficial. He apparently relied on one book, and that written by an exorcist with a conflict of interest. The book is terribly biased,[32] coming as it does from a tradition that developed harsh methods for dealing with dissent. Peck's treatment of the subject is anything but scholarly. Each time he entertains the thought of explaining more to the reader, he resorts to a pious dodge, such as protecting privacy or confidentiality.

Or why did Peck feel it necessary to say that both cases involved *satanic* possession? Did meeting "Satan" somehow enhance his standing or the credibility of speaking of possession? Are we to assume that the judgment about *satanic* possession justifies the assault on the patient?

---

[31]Peck, *Lie*, 201.

[32]Malachi Martin, *Hostage to the Devil: The Possession and Exorcism of Five Americans* (Harper: San Francisco, 1992—orig. NY: Bantam, 1977). 512 pages. The publisher's blurb calls this "a chilling and highly convincing account of possession and exorcism in modern America, hailed by NBC radio as 'one of the most stirring books on the contemporary scene.'"

Another problem pertains to "informed consent." Who signed the form—the possessed patient, or Satan, the one "possessing" her body? Or was it a parent who was badgered and coerced by trusted religious leaders and the patient had no way to judge their profound immersion in superstition?

Furthermore, Peck is contradictory about the exorcisms. He says they "were successful" but that the patients had emotional problems that continued in the weeks afterward.[33] They continued to have delusions and thus remained in psychotherapy.[34] How, then, are the exorcisms in any way to be called successful?

In one of the very few stories of exorcism in the New Testament, Mark says that the man was left "in his right mind" (Mark 5:15). Most writers say that demon-possessed persons are quickly released and cured through exorcisms.[35] Why, then, was Peck's patient left with little change in her mental state?

Further, why would possessions be "rare," as Peck says, if they are actually the work of Satan and his minions? Are there only a few demons in Satan's army, or are they simply sleeping on the job? Are people so powerful or Satan so benign and apathetic that possession is of little interest to them?

Finally, there is a terrible irony in the fact that the other patients Peck described as being "perfectly possessed" manifested none of the features of possession as outlined in the *Rituale Romanum*, upon which he has relied in the chapter on possession.

The most serious problem with the account, however, is that Peck has set aside 200 years of medical science by reverting to demon possession as an explanation for severe mental illness. He goes immediately into the world of the supernatural, looking for evil beings—a quest that has long since been discredited both by theology and science. That may just be its appeal, of course.[36]

In spite of the hype and promises of science and technology, people are still faced with a formidable array of seemingly unsolvable diseases and sociopolitical problems. People are afflicted in body and mind, and science

---

[33]Peck, 189.

[34]Ibid.

[35] John P. Newport, *Demons, Demons, Demons: a Christian Guide Through the Murky Maze of the Occult* (Nashville: Broadman, 1972) 80.

[36] Koch, *Demonology Past and Present* says the increased interest in demonology is related to reactions against rationalism. He sees it as part of the post-modern emphasis, encouraged by anti-modernism and anti-liberalism. See 31, 145.

often seems helpless to offer anything but the hope that "someday" cures will be discovered. The "underside" of the widespread interest in demon possession seems to be a cynical attitude toward or disappointments in science.

Exorcism as a Metaphor for Healing

The proper response to radical suffering is one of recognizing the demonic in human life and experience (cf. ch. 1) without lapsing into a search for little demons or satanic possession. Peck's treatment of a mentally ill teenager as if she were demon possessed shows a lack of professionalism, bad ethics, and uninformed religious perspectives. He shows a terribly superficial understanding of both the biblical materials and the history of demonology. Even more troubling is his endorsement of an exorcism and his attribution of satanic possession to patients he had been treating, but his *use* of the notion of possession as a substitute for psychiatric therapy seems simply incredible. To endorse methods rooted in superstition and fear is to abandon rationality itself and the progress made in scientific medicine.

The Reality of the Demonic

Even so, the concept of the demonic should not be abandoned. There is some truth in speaking of demonic possession and even using the language of exorcism.[37] Even Paul Tillich retained the frame of reference in spite of his project of demythologization.[38]

But our frame of reference and methods of relating to the problem will be quite different than the patterns described in the *Rituale Romanum*, or those attended by psychiatrist Peck. Such approaches lead to the insulting and damaging actions we saw televised by ABC News. What seemed catharsis or relief was apparently from exhaustion or to escape her tormentors, rather than from an exorcism of some indwelling demon. Jenna's mental illness remained, and she continued in therapy.

In addition to all she suffers from mental illness, the church now says she is possessed by demons. If she had problems to begin with, she has even more now that the confrontation with Satan ended in defeat. Televising that discredited ritual makes a mockery of both faith and science, not to mention

[37]Koch himself speaks of the demonic in a way that is adjectival. He apparently intends to retain a degree of ambiguity as to whether he means demonic possession or being "possessed" by demonic ideas, beliefs, etc. See 133ff.

[38]See Paul Tillich, *Dynamics of Faith* (Harper & Brothers, 1957) 48ff., esp. 50.

media integrity. Those associated with it were justifiably subjected to professional criticism and theological ridicule.

The reality of the "demonic" should be combined with the methods of science. Suffering may well be the human experience of the demonic, and healing may be seen as the exorcism of demons. The reverse order of relationships leads to perverse practices: demons are not the source of our sufferings, and exorcisms do not lead to healing. The late Arthur C. McGill seemed to have it right in saying that "demonism constitutes the dominant religious experience of our time...and suffering poses the central theological problem."[39]

People often resort to "devil talk" whenever they confront overwhelming powers at work that elude ordinary explanations. "Devil" and "demon" (Pazuzu, Satan, Beelzebub, Lucifer) are mythological names given to realities that people confront but apparently can neither explain nor control. Using the language of the demonic or Satan is a process of "personification"—naming a problem in order better to communicate the character of the experience itself. Not only is communication facilitated, but the speaker is also given power over the force. To know the name is to gain a measure of control over extraordinary powers. When the name of the "demon" is known, it can be "exorcised."

To refer to a phenomenon as "demonic" or "satanic," or to indicate it is the work of "Satan" or "Pazuzu," is to say that this experience defies explanation in terms of the activity of God who is good, or even of human evil, even though people are sinners. There is evil in the world. People are being destroyed by forces beyond their control. These powers operate at both personal and social levels—fear, hate, greed, corruption, aggression, lust, the list could go on and on. These are the works of "the devil," the enemy of people and a contradiction to the will and intention of God. Their power is real; their destructiveness can be permanent and widespread.

Speaking of the Unspeakable

In the absence of descriptive language, Paul referred to "the mysterious (spiritual) powers of the universe" (Eph 5:12). The struggle is not with human enemies, but with spiritual realities that affect our *persona*, that destroy hope, that ravage mind and body. We face an enemy with a voracious appetite for the innocent and an uncanny ability to wreak havoc

---

[39]Arthur C. McGill, *Suffering: a Test of Theological Method* (Philadelphia: Westminster, 1968) from the Foreword by Paul Ramsey, 10.

with human life, both personally and socially. There would be something terribly unsatisfactory about so cleansing our vocabulary that we cannot even speak of the evils that overwhelm us.

Rejecting the idea of the demonic in the name of objective science may actually make us more susceptible to its influence and power. As Marion Woodman puts it,

> Despite our so-called liberation from gods and demons, few can live without them. Their absence makes nothing better. It may even make everything worse.... The world of the archetype is now an open market for the general populace without any ritual containment. If we are blindly living out an archetype, we are not containing our own life. We are possessed, and possession acts as a magnet on unconscious people in our environment. Everyday life becomes a dangerous world where illusion and reality can be fatally confused.[40]

We need ways to address the depths of our anguish and frustration in the face of forces that destroy and maim us. But resorting to a search for demons and possessing devils diverts attention to the depths of the problem and the self-delusions that cause people to be part of the evil they abhor.

Advocating an exorcism for dealing with a person with severe mental illness poses more problems than it solves. Marc Cramer put it right in saying that "demoniacal possession exists.... The Devil [is] within us all—mentally real, but not physically or metaphysically extant, at least not demonstrably so. [Most if not all] cases brought before the exorcist, anthropologist, mental science researcher or shaman are either hysterias, psychotic manifestations or outright frauds."[41]

Speaking in Parables

Both Scripture and ethics need the language of and ability to speak in parables. The language is figurative. Symbolic language is not to be taken literally, but the reality to which it points is to be taken with extreme seriousness.

---

[40]Marion Woodman, *The Pregnant Virgin: a Process of Psychological Transformation* (Toronto: Inner City Books, 1985) 20.

[41]Marc Cramer, *The Devil Within* (London: W. H. Allen, 1979) 12.

The Gerasene demoniac is a story of healing by Jesus (Mark 5:1–13; Matt 8:28–32) that embodies these truths. It is a story—a parable—of Jesus' confrontation with the demonic in human life. Three truths stand out in the story. First, a man is suffering inexplicably and horribly. Overcome by the sheer magnitude of his problems, he is incapable of breaking his bonds. He mutilates himself and engages in bizarre behavior. Second, his sufferings were interpreted as demonic possession. Even epilepsy was regarded as demon-induced in the first-century world and until modern science discovered and treated it as a neurological disorder, but there was no medicine that was capable of curing the Gerasene. He was overcome by the powers of evil. Third, Christ brought healing to this man by casting out the evils that had overcome him. Exorcism seemed the best metaphor available to the biblical writers.

Science has enabled a great deal of healing of both mind and body. Both medicine and psychiatry have cast out numerous devils: smallpox has been overcome, polio has been defeated, and many of the mentally ill have been made whole. Some are able to function in homes and jobs. At one time, their problems seemed insurmountable, a mystery and source of continuing frustration to their caregivers. Now they rejoice at the healing that has taken place. Did Jesus not say, "You will do greater works than these?" (John 14:12). Good medicine is engaged in "healing the sick…[and] casting out demons."[42]

Poetic imagery permeates stories like that of the demoniac of Gerasene. When speaking of demons and the demonic, we are dealing with the language of poetry and its meanings.[43] Dante, the poet-theologian who helped to create much of the imagery associated with a literal hell, where the fire is unquenchable and damnation is eternal, had a certain message in mind. Both he and Milton wrote with the caveat that "the mind is its own place, and in itself / Can make a heav'n of hell, a hell of heav'n."[44]

---

[42] Tillich, *The Eternal Now*, 59.

[43] Note that Isaiah 14:12 is Hebrew poetry, though it is often taken as *the* passage that explains the origins of Satan, or Lucifer. A precise translation is: "Bright morning star, how you have fallen from heaven, / thrown to earth, prostrate among the nations!" (NEB) The passage is actually referring to the coming humiliation of the King of Babylon (v. 3) and the poetic section is a taunt that the prophet says can be sung against the King.

[44] See Milton, *Paradise Lost*, Bk. I, 1, 253 and Aleghieri Dante, *The Divine Comedy*, trans. John Ciardi (New York: New American Library, 1954). The Princeton Project website gives the complete text of *The Divine Comedy* and the works of Dante in

Peter Blatty's book *The Exorcist* is also an instructive example. It is a parable of good and evil, a story of the power of the demonic in human life. At one level, it is the story of a girl whose inexplicable suffering and violent behavior lead to a diagnosis of demon possession and efforts to cure her through an exorcism. This phenomenon points to a fundamental problem when reading such tales of horror. The vivid imagery employed and the shocking details of suffering and evil that are portrayed cause one to major on the obvious—to "see" the imagery but fail to understand the reality to which the imagery points. This is precisely the "second level" at which I see Blatty's story operating as a parable of the human confrontation with evil.[45]

As a parable, a distinction must be made between what the author is saying and how it is said, or between the message conveyed and the imagery used to convey the message. Blatty had more in mind than simply to produce a story based on a case of exorcism in 1949. The supposed "possession" in that case can neither be proven nor disproven at this point in time. *The Exorcist* goes beyond that, however, using some of the details as elements in a gripping portrayal of the power of the demonic in human life. For Blatty, "the demonic"—that which is destructive of human life and bodily integrity—is a reality to be reckoned with. The problem is how to convey the extent and power of that reality. How does one express the inexpressible or portray "the invisible powers of the universe" (cf. Eph 2:10)?

Poetry is the only adequate form of language to address the reality of evil. It allows fanciful imagery that is not simply one-dimensional, as in prose or descriptive language. People are not able to comprehend or express the incomprehensible and mysterious. Who can speak adequately of the horrors of hell or the inexorable, overwhelming evils of the demonic?

The reality of hell should also not be dismissed. Hell is real. It is "a place prepared for the devil and his angels" (Matt 25:41). Hell is not a people-friendly place—but many people live there. Many know where hell is—they have seen it in the haunted, terrified eyes of the mentally ill; or in the agonies, despair, and hopelessness of the depressed or those unjustly imprisoned or tortured. It has been seen it in the flat affect of the psychotic. No one seems to be at home. An evil resides there.

English and Italian. The *Comedy* in English can be downloaded from Project Gutenberg.

[45]See Paul D. Simmons, "Demonology in *The Exorcist*," *Movies and Ministry*, II:7, March, 1977.

It may be heard in the screams of the grieving—those whose pain is unquenchable, whose hunger for justice is never satisfied, who live with the worms that eat at their minds every waking moment. Why is my child dying of cancer? Read the haunting novel by Peter DeVries about a father's relationship to his eight-year-old daughter who is dying of leukemia.[46]

Why are our children born with Hurler's Syndrome? Why epidermosa bulosa? A child needs the human touch that cannot be given, or that is given at enormous cost. The mother's gentlest, most loving touch causes painful blisters! Or what of those born anencephalic (with no or only a partial brain)? Why is my child so cursed that he will never speak the first word or read the first sentence? It will never know either its parents or its caretakers. Why has this person's humanity been irrevocably distorted in the mysteries of genetic combinations or mutations?

Hell is a place of abject loneliness. Who can enter the realm of the cursed—of those possessed by powers that overwhelm them? This is a lonely place; no one can share it, for they cannot share the misery, the sense of desperation, of abandonment. It is solitary confinement. "Abandon hope all you who enter here," Dante wrote over the gates to the terrible city below. It is a lonely and hopeless place to the mind distorted by powers that forbid organized thought, ordered reflexes, meaningful activity, or sensible conversation. Who can enter this world and share the disorientation and confusion of many voices?

Hell may be a lonely place, but many dwell there. Thousands inhabit this land of God-forsakenness—of brokenness. In this place, the worms of disease and bodily distortion never die. Our hospitals are full and running over of the broken and bewildered. Or they are walking the streets since there is no room in the hospital. Their brokenness is mirrored in the continued and unresolved grief of parents and loved ones; in the concerns and anxieties of those who care for them; of the psychiatrists, social workers, case workers, and legal aides who work with them.

Their caregivers face the formidable tasks of diagnosis, treatment, negotiating with resistant patients, of touching the boundaries of the world in which patients live and others only dimly understand. The human mind—who can know it? The disorders of the mind—what can heal them? The demonic in human experience takes many forms.

How do we capture the frustration of wanting to think clearly and speak distinctly but we cannot? How do we express the horror of discovering that your neurological system is deteriorating and there is neither cure nor

---

[46]Peter DeVries, *The Blood of the Lamb*. (New York: Signet Books, 1961).

treatment? Huntington's disease is a nightmare lurking in the shadows of our genes. Amolytropic sclerosis strikes down the strong and the mighty in the midst of their powers and promising careers. It is ironic, but fitting, that we know it as Lou Gehrig's disease.

These are people who live in hell. It is not a place of their own making. It is called Hurler's Syndrome, Tourette's Syndrome, schizophrenia, Tay Sachs—evils beyond anyone's control. They slaughter the innocent on the rack of cruel torture and unrelenting pain, both physical and psychic. They destroy people slowly or quickly, but always surely and effectively. They are reminders that people are subject to authorities beyond the powers of medical science to control or correct.

What power sends the genetic code into chaos? Why should nucleotides spin out of control into a disastrous replication of normal sequences? The radical disease Fragile X results from extensive replication of a basic sequence of CGG. Angelman Syndrome is a severe form of retardation accompanied by excess laughing and movement. The cause seems a missing bit of DNA on the X chromosome. Such severe distortions are the experience of the demonic in human life. Luther once said that such children are "changelings" created by Satan, not by God.[47] He saw the problem, even if his resolution is hardly satisfactory.

Any destructive illness poses the issue of the demonic. The reality of death can be accepted. Death belongs to the human story because people embody a creaturely existence. Death is a consequence of human finitude—of creaturehood. Mortality is both our dignity and our curse. We "think we were not made to die," said Tennyson, but we are. If we can transcend our finitude by thinking thoughts of eternity, we also distort our importance by thinking we either were or should have been created immortal. The contradiction is in the language—we are *created*, which is to be mortal—to have our lives bounded by birth and death.

Death and the demonic may certainly be related. Death may have demonic dimensions. Uncle *Thanatos* always calls his imps alongside to do his bidding—to create anxiety, to tempt us to think we are immortal, to throw us into panic, to tell us death is unnatural, ungodly, that we should eradicate death from the human scene. But should we? We need only think of the hellish existence to which the world would already have been subjected had death been eradicated or never created. Were every person

---

[47]See Paul Althaus, *The Ethics of Martin Luther* (Fortress, 1972) 96, n. 83, citing *Luther's Works*, 45:396–97.

still alive who had ever been born, earth would be a hell to itself—a nightmare into which only the hapless would be born.

The demonic, then, is not to be found in the ordinary circumstances of life, nor even in the vicissitudes or misfortunes that may cross our paths and alter our course. These we can accept. For them we find grace in history, medicine, and nature. We not only cope but may well thrive on the challenge, the tension with daily struggles that are manageable even if troublesome.

The demonic is in the extraordinary, the cruel, the unconquerable, the unmanageable, the unfathomable. It is in devastating, destructive illnesses that we confront the face of evil. They are legion. In them we experience dark powers beyond us. They baffle both science and religion. They seem beyond the reach of both prayer and medication. It is the experience of powers that work evil in us, that destroy us. They pose for us the protest of the prophet: "Is there no balm in Gilead; is there no physician there? Why has healing not taken place?" (Jer 8:22).

The demonic perhaps touches us most intimately in the derangement of the mind. At least Peck had that much right. Capacities for humanness are severely diminished, if not destroyed. The qualities of empathy, of caring, of conscience, of remorse, and of true insight into self may be inevitably lost to the inexorable forces of synapses that do not work, of neuron transmitters sending garbled messages. The code may be unbreakable, and the problem may appear unsolvable.

Severe mental illness is baffling, bewildering, and terrifying. Such people seem subjected to a type of living death. For this, there is no closure; the grief is ongoing.

Hear the voices; they come from the television; they tell us the FBI and the CIA are after us, or that they want us to work for them, or that the President is making a special trip to see us. They may be from outer space; aliens may be in control, planting transmitters in the brain. The voices give grandiose notions of a mission to and for the world. God has a special job for us to do. It is a secret mission, but indispensable for the salvation of the world or the well-being of society. They give us inside information and extensive knowledge of others and their motives or schemes to dominate or destroy. There seems at times a fine line between those in mental institutions and those who destroy millions with political and military power.

How shall we express the nightmare existence of living with voices that cannot be silenced? They can hardly be trusted, but they must be heard.

They cannot be managed by the exercise of the will, nor controlled by the powers of medication. They are there—from Joan of Arc to David Berkowitz. They are commonplace on the psych ward of the hospital; they talk among themselves. Psychiatrists have names for all this—disassociative reactions, paranoid schizophrenia, grandiosity, thought projection—but the theological question is why should some people be possessed by unfounded fears and anxieties raging out of control? Why these mood swings from extreme agitation to winsome gentleness, from a manic high with near-genius abilities to a destructive depression where life loses all meaning and hope? We are dealing with the demonic in human experience. God would not do this to us, nor is it because the person is evil—they are the victim, not the cause of this evil.

A young man once wrote me a long and poignant letter. He described his "difficulties" as "demonic (satanic) torment and interference." His troubles began after nineteen years in the Air Force, he said, when "satanic (and, later, angelic) spirit beings entered my life (and found) me completely ignorant of the 'real world.'" He found no help from ministers or seminary professors from whom he sought counsel. The Scriptures and the Holy Spirit were unable to deliver him. His "condition/difficulty had been diagnosed officially as paranoid schizophrenia," he said. However, he continued, "[I] *know* that my problem has been severe demonic torment, harassment, and interference. And I feel certain that many (if not most) schizophrenics suffer from the same type of outright attack by demons."[48] Then he added, "I am doing much better now. The demons are still present and still torment me and interfere in my life...but I have been receiving *much* help from the Angelic spirit beings of good will.... They have spoken *aloud* to me...because of the demonic onslaught that I have experienced for so long.... Demons had been speaking *aloud* to me for some time and I was completely baffled, confused, and tormented beyond reason."[49]

In a handwritten note in the margins of an article on schizophrenia[50] included with his letter, he said that all the classical symptoms of schizophrenia had occurred in his life during the past ten plus years. He declared emphatically, however, that he "could *definitely* trace them to demonic attacks, interference, and torment!"

---

[48]Personal letter, p. 6.

[49]Personal letter, p. 7.

[50]P. A. Berger, J. D. Barchas and T. A. Gonda, "Schizophrenia: the Problem and its Treatment," *The Stanford Magazine*, Fall/Winter, 1980, 26–28, 33, 76–77.

The human mind—who can comprehend its mysteries and powers, its capacities, and its almost limitless ability to respond to suggestion? To believe in demons is to create them; to go looking for them is to find them. To tell people they are possessed by demons may be to consign them to the reality of evils they can neither comprehend nor overcome. People create evils by proposing the existence of demons and go probing and prodding to exorcise those planted by suggestion.

Jüngian psychology speaks of "possession," but in a way far different than the advocates of demonism. People often come under the pervasive influence of transpersonal powers. They are us—they are not us. They are our "shadow side," as Jüng put it.[51] They are powers over us and within us. They belong to us, but they may dominate and destroy us.

Stories like *The Exorcist* may also help to restore our awareness of the reality of the demonic, if we do not get stuck on demons or the "reality" of Pazuzu. In that story, "demon possession" is a dramatic device employed to convey the reality of evil in human life. The theme of the power and stubbornness of evil permeates the story. The best of modern science failed to cure the problem; the most educated, enlightened people of the time could not grasp the dimensions of the evil destroying the main character of *The Exorcist*, Regan. The only effective cure was to be found in beginning with a childlike belief in the reality of evil. Once "it" was "named," it could be dealt with; it could be exorcised; she could be healed.

How else might the excesses of Nazi power that destroyed six million Jews and attempted to tyrannize the world be explained? How might the evil of Stalin's extermination of ten million political opponents in Siberia be named? To what power might one ascribe the thoroughgoing racism and reliance on military power and political intrigue that seems to infect our own country? The Bible names it: these are from Satan, the father of all liars and murderers (John 8:44).

Furthermore, evil is difficult to overcome. Neither scientists nor moralists will find it easy to defeat the enemy. Like Father Merrin in *The Exorcist*, we are constantly at war with the Antagonist. All the powers of science and the prayers of the religious will be necessary if the evils that afflict us are to be overcome. We need men and women of compassion,

---

[51] For Jüng the shadow was the sum of all elements that are denied expression and thus becomes an autonomous splinter personality, or autonomous complex. It was not only the source of evil but also of certain good qualities. See C. J. Jüng, G. Adler, and R. F. C. Hull, *The Archetypes and the Collective Unconscious*, Collected Works, 2nd ed. (Princeton: Princeton University Press, 1981) vol. 9, 284.

intelligence, and determination to commit themselves to conquering the mysterious powers that still control and often destroy us.

Conclusion

A two-pronged attack on the diseases of the mind and body is needed. Science and religion are called to a partnership with God. Religion can provide a conceptual frame of reference that helps us know the difference in the good we pursue as the will of God, and the evils we seek to defeat. Science can contribute insights into the nature of the world and the processes of healing—both physical and mental—that offer hope for wholeness instead of brokenness and disease. It can also help to keep our minds focused on the issue and not become engaged in a futile search for metaphysical entities that do not exist.

Tillich was right. Healing can be seen as exorcism. Radical illness is not a matter of demonic possession. Of that we can be certain on both biblical and scientific grounds. The images of "demon possession" and "exorcisms" should be rejected in the name of both good religion and good medicine. We do not begin with the mythological and adapt the scientific process. We begin with scientific understandings and adapt them to the realities and difficulties associated with achieving human wholeness.

Scott Peck got it right when he said that the "biggest step in the healing process...occurs when the client first decides to see a psychotherapist."[52] Had Peck listened more carefully to one of his patients, he would have avoided the misguided path he took into looking for Satan and engaging in exorcisms. As she said to him, "All psychotherapy is a kind of exorcism!"[53]

The exorcism broadcast by ABC television was an incredible display of superstition parading as religious truth and a perversion of medicine into medieval quackery. Both the church and religious faith were understandably subjected to ridicule as a result of that showing. The "exorcism" had no redeeming qualities; it was a setback for the patient, religion, and medical science. In addition to her chronic illness, Jenna now carries the humiliating label of being "possessed by demons," and apparently incurably so. In effect, the church declared that even God is incapable of casting out her demons. Medical science needs to distance itself from such charades both for the sake of the truth of God and the integrity of medicine.

---

[52]Peck, 191.
[53]Peck, 185.

The new partnership between science and religion is not to get both clergy and physicians involved in rituals of exorcism. Both should reject the notions of demons and focus energies on the demonic. There are evils enough to engage the best of our talents, energies, and prayers. Helping patients achieve health and wholeness can be accomplished by consecrated discipline and the vigorous pursuit of the keys to unlock those mysterious powers that still ravage our minds and bodies. Illness confronts us with the reality of the demonic in human experience. Both religion and science are to be engaged in the challenge Jesus delivered to his disciples: "Heal the sick...cast out demons."

ACKNOWLEDGMENTS

Whatever in these pages is helpful, truthful, or insightful is because of the assistance I have received from many people. Their generous commitment of time to read and reflect on the manuscripts and their compassionate encouragement in the midst of constant interruptions and frustrations have been critical factors in bringing this book to completion.

Dr. Richard Clover first made it possible and then encouraged my work as a theologian-ethicist in the context of medical school education. The experience has been profoundly rewarding at a personal level and a serious challenge for professional development.

I am especially grateful to Dr. James O'Brien, chair of the Department of Family and Geriatric Medicine, for making it possible for me to devote time to the preparation of this manuscript. He was also an important part of the chapter on aging and dignity and shared in an earlier essay that was published as "Ethics & Aging: Confronting Abuse and Self-Neglect" in the *Journal of Elder Abuse and Neglect* 11/2 (Winter 2000): 33–54.

Dr. David Doukas, now head of the program in Ethics and Professionalism in Medical Education, has also given constant encouragement to integrate religious and philosophical perspectives into the four-year curriculum of the medical school. Dr. Hiram Polk graciously welcomed me to surgery rounds and helped me think critically about the role of hope in the ICU.

A special word of appreciation belongs to Dr. George Pantalos, who now heads the artificial heart program at Jewish Hospital. He has been a friend and mentor, and I acknowledge special gratitude for his reading and criticism of the chapter on cyborgs. Dr. Gordon Tobin has helped me think critically and realistically about the ethics of hand and face transplants. Dr. Osborne Wiggins has given generously of his time to read manuscripts and help them on to completion. He has been a constant support for my work in the department of philosophy. I also thank Dr. John Barker, who first invited me to contribute an essay on "Ethical Considerations in Composite Tissue Allotransplantation (CTA)," *Microsurgery* (2000), 20:458–465, which was an earlier version of the chapter in this book.

Dr. Keith Parker, a specialist in Jüngian psychology and director of the Connestee Counseling Center, Brevard, North Carolina, was especially helpful with the chapter on exorcism.

I am also indebted to the faculty and administration of Gardner-Webb University (NC) and Campbellsville University (KY) for their invitation to address issues in the right to die, physician-assisted suicide, and embryo stem-cell research. I learned a great deal from their insights and responses, as I did from Dr. David McKenzie and others at Berry College (GA) who generously provided a forum for discussions of abortion and method in ethics.

I also appreciate the cooperation from Common Ground Publishers, who had the right to first publication for my essay "The Artificial Heart: Cyborgs and the Human Future," which appeared in *The International Journal of the Humanities* 2/3 (2006).

A special word of appreciation also goes to the receptionists and support staff of the Department of Family and Geriatric Medicine. Helping me past the technical glitches of the computer age has often fallen to them and the fine staff at IT.

My deep gratitude goes to the Wellcome Library, London, England, and its helpful staff for assistance with the use of the portrait by Balbases that graces the cover of this book. I am also grateful to the American Medical Association for the use of the principles of medical ethics found in Appendix II and the quotation that helps to focus chapter 1 on the purposes of medicine. Thanks also to the University of Virginia Press, Charlottesville, for the use of comments by Thomas Jefferson, without whose wisdom no book can be complete. The Academy of American Poets permitted me to use a portion of "Go Down, Death," a remarkable and moving poem/sermon from the book by James Weldon Johnson, *God's Trombones*, first published in 1927 by the Viking Press. Finally, my gratitude goes to the Vatican for graciously allowing me to summarize passages from *Respect for Life*, which appear as Appendix V.

Appendix I

AUTHORITY IN CLINICAL DECISION MAKING

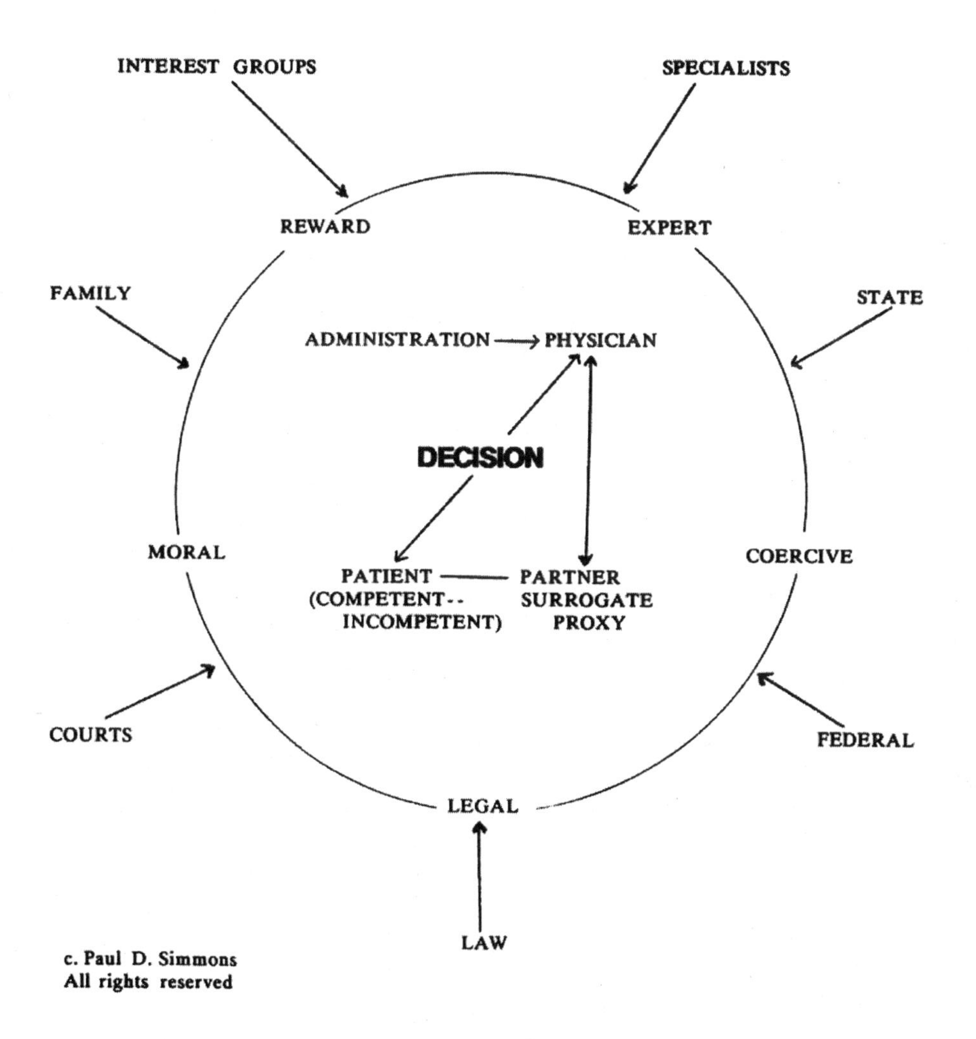

Appendix II

American Medical Association Principles of Medical Ethics

Preamble

The medical profession has long subscribed to a body of ethical statements developed primarily or the benefit of the patient. As a member of this profession, a physician must recognize responsibility to patients first and foremost, as well as to society, to other health professionals, and to self. The following Principles adopted by the American Medical Association are not laws, but standards of conduct which define the essentials of honorable behavior for the physician.

I. A physician shall be dedicated to providing competent medical service with compassion and respect for human dignity.

II. A physician shall uphold the standards of professionalism, be honest in all professional interactions, and strive to report physicians deficient in character or competence, or engaging in fraud or deception, to appropriate entities.

III. A physician shall respect the law and also recognize a responsibility to seek changes in those requirements which are contrary to the best interests of the patient.

IV. A physician shall respect the rights of patients, of colleagues, and of other health professionals, and shall safeguard patient confidences and privacy within the constraints of the law.

V. A physician shall continue to study, apply and advance scientific knowledge, maintain a commitment to medical education, make relevant information available to patients, colleagues, and the public, obtain consultation, and use the talents of other health professionals when indicated.

VI. A physician shall, in the provision of appropriate patient care, except in emergencies, be free to choose whom to serve, with whom to associate, and the environment in which to provide medical care.

VII. A physician shall recognize a responsibility to participate in activities contributing to the improvement of the community and the betterment of public health.

VIII. A physician shall, while caring for a patient, regard responsibility to the patient as paramount.

IX. A physician shall support access to medical care for all people.

Appendix III

The Hippocratic Oath

I swear by Apollo Physician and Hygieia and Panaceia and all the gods and goddesses, making them my witnesses, that I will fulfill according to my ability and judgment this oath and this covenant: To hold him who has taught me this art as equal to my parents and to live my life in partnership with him, and if he is in need of money to give him a share of mine, and to regard his offspring as equal to my brothers in male lineage and to teach them this art—if they desire to learn it—without fee and covenant; to give a share of precepts and oral instruction and all the other learning to my sons and the sons of him who has instructed me and to pupils who have signed the covenant and have taken an oath according to the medical law, but to no one else.

I will apply dietetic measure for the benefit of the sick according to my ability and judgment; I will keep them from harm and injustice.

I will neither give a deadly drug to anybody if asked for it, nor will I make a suggestion to this effect.

Similarly, I will not give to a woman an abortive remedy. In purity and holiness I will guard my life and my art.

I will not use the knife, not even on sufferers from stone, but will withdraw in favor of such men as are engaged in this work.

I will refuse to treat those who are overmastered by their diseases, realizing in such cases that medicine is powerless"

What I may see or hear in the course of the treatment or even outside of the treatment in regard to the life of men, which on no account one must spread abroad, I will keep to myself holding such things shameful to be spoken about.

If I fulfill this oath and do not violate it, may it be granted to me to enjoy life and art, being honored with fame among all men for all time to come; if I transgress it and swear falsely, may the opposite of all this be my lot.

APPENDIX IV

THE NUREMBERG CODE[1]

1. The voluntary consent of the human subject is absolutely essential. This means that the person involved should have legal capacity to give consent; should be so situated as to be able to exercise free power of choice, without the intervention of any element of force, fraud, deceit, duress, over-reaching, or other ulterior form of constraint or coercion; and should have sufficient knowledge and comprehension of the elements of the subject matter involved, as to enable him to make an understanding and enlightened decision. This latter element requires that, before the acceptance of an affirmative decision by the experimental subject, there should be made known to him the nature, duration, and purpose of the experiment; the method and means by which it is to be conducted; all inconveniences and hazards reasonably to be expected; and the effects upon his health or person, which may possibly come from his participation in the experiment.

The duty and responsibility for ascertaining the quality of the consent rests upon each individual who initiates, directs or engages in the experiment. It is a personal duty and responsibility which may not be delegated to another with impunity.

2. The experiment should be such as to yield fruitful results for the good of society, unprocurable by other methods or means of study, and not random and unnecessary in nature.

3. The experiment should be so designed and based on the results of animal experimentation and a knowledge of the natural history of the disease or other problem under study, that the anticipated results will justify the performance of the experiment.

4. The experiment should be so conducted as to avoid all unnecessary physical and mental suffering and injury.

---

[1] "*Trials of War Criminals before the Nuremberg Military Tribunals under Control Council Law No. 10*", Vol. 2, pp. 181–82. Washington, DC: US Government Printing Office, 1949.

5. No experiment should be conducted, where there is an *a priori* reason to believe that death or disabling injury will occur; except, perhaps, in those experiments where the experimental physicians also serve as subjects.

6. The degree of risk to be taken should never exceed that determined by the humanitarian importance of the problem to be solved by the experiment.

7. Proper preparations should be made and adequate facilities provided to protect the experimental subject against even remote possibilities of injury, disability, or death.

8. The experiment should be conducted only by scientifically qualified persons. The highest degree of skill and care should be required through all stages of the experiment of those who conduct or engage in the experiment.

9. During the course of the experiment, the human subject should be at liberty to bring the experiment to an end, if he has reached the physical or mental state, where continuation of the experiment seemed to him to be impossible.

10. During the course of the experiment, the scientist in charge must be prepared to terminate the experiment at any stage, if he has probable cause to believe, in the exercise of the good faith, superior skill and careful judgment required of him, that a continuation of the experiment is likely to result in injury, disability, or death to the experimental subject.

APPENDIX V

CONTRACEPTION AND IN VITRO FERTILIZATION: THE VATICAN RESPONSE

A Summary of: The Vatican, *Instruction on Respect for Human Life in its Origin and on the Dignity of Procreation: Replies to Certain Questions of the Day*[1]

Fundamental principle: "the dignity of the person and his or her integral vocation." The criteria of moral judgment are "the respect, defense and promotion of man, his 'primary and fundamental right' to life, his dignity as a person who is endowed with a spiritual soul and with moral responsibility and who is called to beatific communion with God."

3. Anthropology and Procedures in the Biomedical Field. "The natural moral law expresses and lays down the purposes, rights and duties which are based upon the bodily and spiritual nature of the human person....not simply a set of norms on the biological level, but defined as the rational order..." special concern with procedures which are not strictly therapeutic, e.g. those aimed at improvement of the human biological condition."

4. Fundamental Criteria for a Moral Judgment. "The inviolability of the innocent human being's right to life from the moment of conception until death..."

5. Teachings of the Magisterium. "...life must be actualized in marriage through the specific and exclusive acts of husband and wife, in accordance with the laws inscribed in their persons and in their union."

Part I.

1. Respect due the human embryo: "must be respected—as a person—from the very first instant of his existence."..."Life once conceived must be protected with the utmost care; abortion and infanticide are abominable crimes" (Vat. II)

---

[1] Issued Tues. March 10, 1987, published in 1988 by Congregation for the Doctrine of the Faith. Printed in its entirety in *The New York Times*, 11 March 1987, 10ff.

2. Prenatal Diagnosis (is) morally licit if purpose is to safeguard the life and integrity of the embryo and the mother, without subjecting them to disproportionate risks. ...gravely opposed to moral law if done with thought of possibly inducing an abortion depending upon the results (as malformation or genetic illness).

3. Therapeutic procedures on embryo morally licit? yes, if directed toward its healing and promotion of well-being w/o doing harm to its well-being

4. Moral Evaluation of Research/Experimentation on Embryos (proceeds with the) same criteria as on adult persons; if not directly therapeutic it is illicit; there is no justification for experimentation on embryos or fetuses in or outside the womb for noble objectives....(it is a) crime against their dignity.

5. Use of embryos in vitro: (Embryos are) human beings with rights; immoral to produce knowing they to be exploited as disposable biological material. ...duty to condemn the particular gravity of the voluntary destruction of human embryos obtained ....for the sole purpose of research, either by means of artificial insemination or 'twin fission'" (i.e. cloning; ed.).

a. usurps the place of God.

b. sets self up as master of the destiny of others –

c. arbitrarily chooses who will live who will be sent to death.

6. Other procedures: contrary to the human dignity proper to the embryo, ...to right of every person to be born within marriage and from marriage; ...twin fission, cloning or parthenogesesis are contrary to the moral law; cryopreservation constitutes an offence against the respect due to human beings; as are attempts to influence chromasomic or genetic inheritance—not therapeutic but aimed at producing selected by certain qualities.

Part II. Interventions Upon Human Procreation.

Artificial fertilization—refers to technical procedures to obtain conception other than through sexual union of a man and woman.

IVF and AI—Practice of IVF has resulted in destruction of innumerable embryos...leads to man's domination over the life and death of his fellow human beings and can lead to a system of radical eugenics.

Heterologous (AID)—procreation must be the fruit of marriage; "fidelity of the spouses...involves a reciprocal respect of their right to become a father and a mother only through each other." Artificial Fertilization of woman who is unmarried or a widow whoever the donor may be cannot be morally justified.

Surrogate motherhood—not morally licit;—objective failure to meet the obligations of maternal love, conjugal fidelity and responsible motherhood;

B. Homologous (AIH)—IVF, ET, and AIH—

4. connection between procreation and conjugal act? *contraception is forbidden* as contrary to the unitive and procreational aspects of conjugal acts; conception must be desired as the fruit of the conjugal act. sexual intercourse must take place with openness to procreation; the fruit and result of married love;

5. *AIH in vitro fertilization:* desire for a child is good intention but not sufficient to justify IVF between spouses. IVF is not related to conjugal act; (de facto not enough); cannot be admitted "except for those cases in which the technological means is not a substitute for the conjugal act but serves to facilitate and to help so that the act attains its natural purpose."... *masturbation* (to obtain semen) deprived of unitive meaning...

7. The moral criterion: medical intervention is good if it seeks to assist the conjugal act.

8. *Suffering* of infertility: marriage does not confer the right to have a child; only the right to engage in acts which are ordered to procreation; a child is not an object to which one has a right; a child is a gift; right to be respected as person from conception.

Infertility as cruciform existence; share the Cross, way of spiritual fruitfulness. Sterility the occasion for fruitful service (adoption, educational work, assist families and handicapped);

Part III: Moral and Civil Law

Legislation: cannot rely on conscience of individuals or self-regulation of researchers; "eugenism" is deplorable; legislation should prohibit embryo banks, post mortem insemination and surrogate motherhood. "...the duty of the public authority to insure that the civil law is regulated according to the fundamental norms of the moral law in matters concerning human rights, human life and the institution of the family.... legislation (often) "confers an undue legitimation of certain practices"; "seen as incapable of guaranteeing that morality which is in conformity with the natural exigencies of the human person and with the 'unwritten laws' etched by the Creator upon the human heart."

Joseph Card. Ratzinger, Prefect

APPENDIX VI

PHYSICIAN-ASSISTED SUICIDE: *VACCO V. QUILL*[1]

Chief Justice William Rehnquist delivered the opinion of the court:

Whether New York's law against assisted suicide violates the Equal Protection Clause of the Fourteenth Amendment. We hold that it does not. (Under New York law, a person is guilty of manslaughter in the 2nd degree when...he intentionally causes or aids another person to commit suicide...a class C felony.)

Equal Protection Clause requires that no person be denied the equal protection of the laws. "This provision creates no substantive rights. Instead, it embodies a general rule that States must treat like cases alike but may treat unlike cases accordingly. If (a law) neither burdens a fundamental right nor targets a suspect class, we will uphold [it] so long as it bears a rational relation to some legitimate end."....

Neither the NY ban...on assisting suicide nor its statutes permitting patients to refuse medical treatment treat anyone differently than anyone else or draw any distinctions between persons. *Everyone*, regardless of physical condition, is entitled, if competent, to refuse unwanted lifesaving medical treatment; *no one* is permitted to assist a suicide. Generally speaking, laws that apply evenhandedly to all 'unquestionably comply' with the Equal Protection Clause."...

We think the distinction between assisting suicide and withdrawing life- sustaining treatment, a distinction widely recognized and endorsed in the medical profession and in our legal traditions, is both important and logical; it is certainly rational."

This distinction comports with fundamental legal principles of causation and intent.

---

[1] *Summary statements by Paul D. Simmons taken from: Vacco, Attorney General of New York, et al. v. Quill et al.* argued January 8, 1997, decided June 26, 1997.

[p. 5]...The law has long used actors' intent or purpose to distinguish between two acts that may have the same result....

In affirming the right to refuse treatment, "the state has neither endorsed a general right to 'hasten death' nor approved physician-assisted suicide. The State has reaffirmed the line between 'killing' and 'letting die'." [p. 10]

Court disagrees with the "claim that the distinction between refusing lifesaving medical treatment and assisted suicide is 'arbitrary' and 'irrational'." [p. 11]

(NY Task Force, *When Death is Sought*, "It is widely recognized that the provision of pain medication is ethically and professionally acceptable even when the treatment may hasten the patients' death, if the medication is intended to alleviate pain and severe discomfort, not to cause death." note 11, p. 12)

(Does not mean that "in all cases there will in fact be a significant difference between the intent of the physicians, the patients or the families" in cases of withdrawal of treatment and physician-assisted-suicide cases. note 12, p. 12.)

Justice Stevens' Concurring Opinion

[T]here is also room for further debate about the limits that the Constitution places on the power of the States to punish the practice.

History and tradition provide ample support for refusing to recognize an open-ended constitutional right to commit suicide. Much more than the State's paternalistic interest in protecting the individual from the irrevocable consequences of an ill-advised decision motivated by temporary concerns is at stake. [Citing John Donne's "no man is an island" (p. 4)]

[The right to refuse treatment] was not deduced from abstract concepts of personal autonomy. Instead, it was supported by the common-law tradition protecting the individual's general right to refuse unwanted medical treatment."

APPENDIX VII

SCHIAVO TIMELINE: POLITICS AND THE RIGHT TO DIE

February 25, 1990 Terri Schiavo collapsed in her home from possible potassium imbalance related to anorexia; heart stopped and brain damage resulted from lack of oxygen

June 1990 Nancy Cruzan case decided by Supreme Court

July 29, 1993 Terri's parents, Bob and Mary Schindler, try to have Terri's husband, Michael Schiavo, removed as Terri's guardian. Case dismissed

February 11, 2000 Circuit Judge George Greer approves Michael Schiavo's request to disconnect Terri's feeding tube

April 2001 State and US high courts refuse to intervene, and tube is removed; judge orders tube reinserted two days later

November 22, 2002 After medical testimony, Greer finds no evidence that Terri has hope of recovery and orders tube removed

October 15, 2003 Tube removed for second time

October 21, 2003 Florida governor Jeb Bush signs a hastily passed bill and orders tube reinserted

December 2, 2003 Independent guardian *ad litem* finds "no reasonable medical hope" that Terri will improve

September 23, 2004 Florida Supreme Court rules "Terri's Law" is unconstitutional

February 25, 2005 Judge Greer gives permission for tube removal at 1 P.M., March 18

March 16, 2005 US House passes bill aimed at keeping Terri Schiavo alive

March 17, 2005 Florida House passes bill intended to keep Terri alive; US Senate considers bill different from House version

March 18, 2005 Feeding tube removed—third time for Terri

March 19, 2005 Leaders from both parties agree on a bill that would allow tube to be reconnected while federal judge reviews the case

March 20, 2005 Senate passes bill (S.686); President Bush flies in from Texas to sign the bill, known as the Palm Sunday Compromise

March 21, 2005 House passes the bill, and Bush signs it; Terri's parents file an emergency request with Tampa federal judge to have tube reconnected

March 22, 2005 US District Judge James Whittemore refuses to order reinsertion of the tube; parents appeal to 11th US Circuit Court of Appeals in Atlanta

March 23, 2005 The 11th Circuit Court declines to order reinsertion of tube

March 24, 2005 US Supreme Court refuses to hear the case or to order tube reinserted (cf. court decision of June 1990 in Cruzan case)

March 26, 2005 Permission to give Easter Communion to Terri is denied (she cannot swallow)

March 27, 2005 (Easter Sunday) Attorneys say Schiavo fight nears end

Terri Schiavo died thirteen days after the gastronomic tube was removed.

BIBLIOGRAPHY

Books

Almond, Gabriel, R. Scott Appleby, and Emmanuel Sivan. *Strong Religion: The Rise of Fundamentalisms around the World.* Chicago: University of Chicago Press, 2003.

Althaus, Paul. *The Ethics of Martin Luther.* Minneapolis: Fortress Press, 1972.

Arkes, H. *Natural Rights and the Right to Choose.* Cambridge: University Press, 2002.

Arras, J. D. and B. Steinbock, eds. *Ethical Issues in Modern Medicine*, 5th ed. Guilford, CT: Mayfield, 1999.

Bainton, Roland. *Christian Attitudes toward War and Peace.* New York: Abingdon Press, 1980.

Balmer, Randall. *Thy Kingdom Come: An Evangelical's Lament.* New York: Basic Books, 2006.

Beauchamp, Thomas L. and James F. Childress. *Principles of Biomedical Ethics*, 5th ed. New York: Oxford University Press, 2001.

Becker, Ernst. *The Denial of Death.* New York: Free Press, 1973.

Benson, H. *Timeless Healing.* New York: Simon & Schuster, 1996.

Berry, Wendell. *Another Turn of the Crank.* Washington, DC: Counterpoint Books, 1995.

Blatty, Peter. *The Exorcist.* New York: HarperCollins, 1971.

Bok, Sissela. "Who Shall Count as a Human Being?" In *Abortion: Pro and Con.* Ed. R. C. Perkins. Cambridge: Schenkman Publishing Co., 1974.

Brooks, Rodney. *Flesh and Machines: How Robots will Change Us.* New York: Pantheon, 2002.

Browning, Robert. "Rabbi ben Ezra." In D. H. S. Nicholson and A. H. E. Lee, editors, *The Oxford Book of English Mystical Verse.* Oxford: Oxford University Press, 1917.

Burchhardt, T. *Alchemy: Science of the Cosmos, Science of the Soul.* Trans. W. Stoddart. London: Stuart and Watkins, 1967.

Burman, Edward. *The Inquisition: Hammer of Heresy.* New York: Dorset Press, 1992.

Caldicott, Helen. *Missile Envy: The Arms Race & Nuclear War*. New York: William Morrow, 1984.

Callahan, Daniel. *Setting Limits: Medical Goals in an Aging Society*. New York: Simon & Schuster, 1987.

———. *False Hopes: Why America's Quest for Perfect Health Is a Recipe for Failure*. New York: Simon & Schuster, 1998.

———. "The Immorality of Assisted Suicide." In *Physician-assisted Suicide*. Ed. R. Weir. Bloomington: Indiana University Press, 1997.

Cantwell v. Connecticut, 60 S. Ct. 900 (1940).

Caplan, Arthur L., ed. *When Medicine Went Mad: Bioethics and the Holocaust*. New York: Humana Press, 1992.

Carroll, James. *Constantine's Sword: The Church and the Jews*. New York: Houghton-Mifflin, 2001.

Cassell, Eric J. *The Nature of Suffering and the Goals of Medicine*. New York: Oxford University Press, 1991.

Cole, H. A., with M. M. Jablow. *One in a Million*. Boston: Little, Brown & Co., 1990.

Congregation for the Doctrine of the Faith. *Instruction on Respect for Human Life in Its Origin and on the Dignity of Procreation: Replies to Certain Questions of the_Day*. Rome: The Vatican, 1988.

Conners, R. B. and M. J. Smith. "Religious Insistence on Medical Treatment." *HCR* 26/4 (1996).

Cosgrove, Charles H. *Appealing to Scripture in Moral Debate*. Grand Rapids, MI: Eerdmans Publishing Co., 2002.

Council on Ethical and Judicial Affairs. *Code of Medical Ethics: Current Opinions with Annotations 2006–2007*. Chicago: American Medical Association.

Cramer, Marc. *The Devil Within*. London: W. H. Allen, 1979.

Cruzan v. Director, Missouri Department of Health. US 88-1503, June 1990.

Dante, A. *The Divine Comedy*. Translated by John Ciardi. New York: New American Library, 1954.

Davis, Tom. *Sacred Work: Planned Parenthood and Its Clergy Alliances*. Rutgers, NJ: Rutgers University Press, 2005.

DeVries, Peter. *The Blood of the Lamb*. New York: Signet Books, 1961.

Doerr, Edd and James W. Prescott, eds. *Abortion Rights and Fetal Personhood*. Long Beach, CA: Centerline Press, 1989.

Dossey, Larry K. *Healing Words: The Power of Prayer and the Practice of Medicine*. New York: HarperCollins, 1993.

Draper, James T. *The Conscience of a Nation*. Nashville: Broadman, 1983.

Ellul, Jacques. *To Will and To Do*. Trans. C. Edward Hopkin. Philadelphia: Pilgrim Press, 1969.

———. *The Technological Society*. Trans. John Wilkinson. New York: Vantage Books, 1964.

Falwell, Jerry. *Listen, America*. New York: Doubleday, 1980.

Farley, Wendy. *Tragic Vision and Divine Compassion: A Contemporary Theodicy*. Louisville: Westminster/John Knox, 1990.

Faulkner, William. *As I Lay Dying*. New York: Vintage International, 1930.

Frankl, Victor. *Man's Search for Meaning*, rev. ed. Trans. I. Lasch. New York: Simon & Schuster, 1962.

Fromm, Eric. *To Have or To Be?* New York: Bantam, 1976.

Fukuyama, Francis. *Our Post-human Future: Consequences of the Biotechnology Revolution*. New York: Farrar, Straus and Giroux, 2002.

Garrett, T. M., H. W. Baillie, and R. M. Garrett. *Healthcare Ethics: Principles and Problems*, 4th ed. Columbus, OH: Prentice-Hall, 2001.

Gordon, P. *No More Dying*. Austin, TX: Learning Concepts, 1976.

Groopman, Jerome. *The Anatomy of Hope: How People Prevail in the Face of Illness*. New York: Random House, 2004.

Gustafson, James M. *Ethics from a Theocentric Perspective*. Theology and Ethics, vol. 1. Chicago: University of Chicago Press, 1981.

Hare, J. B., ed. Internet Sacred Text Archive. "Scriptures of All World Religions, Texts on Mythology, Folklore, Traditions, the Esoteric." CD-ROM, 1998–2005.

Harrington, Alan. *The Immortalist*. New York: Random House, 1969.

Harrison, Beverly W. *Our Right to Choose*. Boston: Beacon Press, 1983.

Hauerwas, Stanley J. *Naming the Silences: God, Medicine and the Problem of Suffering*. Grand Rapids, MI: Eerdmans Publishing, 1990.

———. "Reflections on Suffering, Death and Medicine." In *Suffering Presence: Theological Reflections on Medicine, the Mentally Handicapped, and the Church*. South Bend, IN: University of Notre Dame Press, 1986.

Hirshberg, C. and M. I. Barasch. *Remarkable Recovery*. New York: Riverhead Books, 1995.

Hitt, Jack. "Pro-life Nation." *The New York Times*.

Jefferson, Thomas. *Jefferson Himself*. Ed. Bernard Mayo. Charlottesville: University of Virginia Press, 1942.

Jones, David A. *The Soul of the Embryo*. London: Continuum Press, 2004.

Jonsen, Albert, M. Siegler, and W. Winslade. *Clinical Ethics*, 6th ed. Guilford, CT: McGraw-Hill, 2006.

Jonsen, Albert and Stephen Toulmin. *The Abuse of Casuistry: A History of Moral Reasoning.* Berkeley, CA: University of California Press, 1988.

Jüng, Carl J., G. Adler, R. F. C. Hull. *The Archetypes and the Collective Unconscious.* Collected Works, vol. 9, 2nd edition. Princeton: Princeton University Press, 1981.

Kahn, Herman. *Thinking about the Unthinkable.* New York: Simon & Schuster, 1984.

Kaplan, Esther. *With God on Their Side.* New York: The New Press, 2005.

Kimball, Charles. *When Religion Becomes Evil: Five Warning Signs.* San Francisco: HarperSanFrancisco, 2002.

Kischer, W. C. and D. N. Irving. *The Human Development Hoax: Time to Tell the Truth*, 1st ed. Clinton Township, MI: Gold Leaf Press, 1995.

Koch, Kurt. *Demonology Past and Present.* Grand Rapids, MI: Kregel Publications, 1972.

Koop, C. Everett and Timothy Johnson. *Let's Talk: An Honest Conversation on Critical Issues.* San Francisco: HarperCollins, 1992.

Kurzweil, Ray. *The Age of Spiritual Machines.* New York: Viking Press, 1999.

Lamm, Richard D. "Doctors Have Patients, Governors Have citizens." In *Narrative Matters: The Power of the Personal Essay in Health Policy.* Ed. F. Mullan, E. Ficklen, and K. Rubin. Baltimore: The Johns Hopkins University Press, 2007.

Latourette, Kenneth S. *A History of Christianity.* New York: Harper and Brothers, 1953.

LeBow, R. *Healthcare Meltdown: Confronting the Myths and Fixing our Failing System.* Chambersburg, PA: Hood Publishing, 2003.

Letter to Diognetus. In C. C. Richardson, editor, *Early Christian Fathers.* Philadelphia: Westminster Press, 1953.

Levine, Carole, ed. *Taking Sides: Clashing Views on Controversial Bioethical Issues*, 7th ed. Guilford, CT: Dushkin Publishing Co., 1997.

Lewis, M. *Aging in America: Trials and Triumphs.* Westport, CT: Americana Healthcare Corps, n.d.

MacKinnon, Barbara. *Ethics: Theory and Contemporary Issues*, 2d ed. Belmont, CA: Wadsworth, 1998.

Martin, Malachi. *Hostage to the Devil: The Possession and Exorcism of Five Americans.* San Francisco: Harper, 1992.

Masello, Robert. *Fallen Angels...and Spirits of the Dark.* New York: Berkley Publishing, 1994.

McGill, Arthur C. *Suffering: A Test of Theological Method.* Philadelphia: Westminster Press, 1968.

Milton, John. *Paradise Lost*. Book 2. Edited by Thomas Newton. London: Proprietors, 1795.

Moltmann, Jürgen. *Religion, Revolution and the Future*. Trans. Douglas Meeks. New York: Scribner's, 1969

———. *The Experiment Hope*. Trans. and ed. Douglas Meeks. Minneapolis: Fortress Press, 1975.

———. *The Church in the Power of the Spirit*. Trans. Margaret Kohl. San Francisco: Harper & Row, 1977.

———. *Theology of Hope*. Trans. James W. Leitch. New York: Harper & Row, 1967.

———. *Science and Wisdom*. Trans. Margaret Kohl. Minneapolis: Fortress Press, 2003.

Moore, Frances D. *A Miracle and a Privilege: Recounting a Half-century of Surgical Advance*. Washington, DC: Joseph Henry Press, 1995.

Mullan, Fitzhugh, Ellen Ficklen, and Kyna Rubin, eds. *Narrative Matters: The Power of the Personal Essay in Health Policy*. Baltimore: The Johns Hopkins University Press, 2006.

Newport, John P. Demons, Demons, Demons: A Christian Guide through the Murky Maze of the Occult. Nashville: Broadman Press, 1972.

Nietszche, Frederich. *Human All Too Human: On the History of Moral Sentiments*. Stuttgart: Kröner Verlag, 1959. Aphorism 88.

Noonan, John T. Jr., ed. *The Morality of Abortion: Legal and Historical Perspectives*. Cambridge, MA: Harvard University Press, 1970.

Nozick, Robert. "Distributive Justice." In *Anarchy, State and Utopia*. New York: Basic Books, 1974.

Nuland, Sherwin B. *How We Die: Reflections on Life's Final Chapter*. New York: Alfred A. Knopf, 1994.

Ogletree, Thomas. *Hospitality to the Stranger: Dimensions of Moral Understanding*. Minneapolis: Fortress Press, 1985.

Orentlicher, David. "Physician-assisted Dying: The Conflict with Fundamental Principles of American Law." In *Medicine Unbound: The Human Body and the Limitations of Medical Intervention*. Ed. Robert H. Blank and A. L Bonnicksen. New York: Columbia University Press, 1994.

Pagels, Elaine. *The Origin of Satan*. New York: Random House, 1995.

Peck, M. Scott. *People of the Lie: The Hope for Healing Human Evil*. New York: Simon & Schuster, 1985.

Pellegrino, Edmund. "Euthanasia as a Distortion of the Healing Relationship." In *Contemporary Issues in Bioethics*, 4th ed. Ed. T. Beauchamp and L. Walters. Belmont, CA: Wadsworth, 1994.

Pence, Gregory E. *Classic Cases in Medical Ethics*, 4th ed. Boston: McGraw-Hill, 2004.

Peters, Ted. *Playing God? Genetic Determinism and Human Freedom*, 2d ed. New York and London: Routledge Press, 2003.

Phillips, Kevin. *American Theocracy: The Perils and Politics of Radical Religion, Oil, and Borrowed Money in the 21st Century*. New York: Viking/Penguin Press, 2006.

Planned Parenthood of Southeastern Pennsylvania v. Casey, 947 F. 2d 682 (3rd Cir. 1991), 112 S. Ct. (1992).

Pope, Alexander. *An Essay on Man*. In *The Complete Poetical Works of Alexander Pope*. Edited by N. W. Boynton. Boston: Houghton Mifflin Co., 1903. Epistle 1, line 95.

Pope John Paul II. *The Gospel of Life*. New York: Times Books, 1995.

Pope Innocent VIII. *Malleus Maleficarum*. Rome: The Vatican, 1484.

Prescott, James W. "Personality Profiles of 'Pro-choice' and 'Anti-choice' Individuals and Cultures." In *Abortion Rights and Fetal Personhood*. Ed. Edd Doerr and James W. Prescott. Long Beach, CA: Centerline Press, 1989.

"President's Commission for the Study of Ethical Problems in Medicine and Biomedical and Behavioral Research." In *Securing Access to Health Care*, vol. 1. Washington, DC: US Government Printing Office, 1983.

Ramsey, Paul. *Fabricated Man: The Ethics of Genetic Control*. New Haven: Yale University Press, 1970.

———. *The Just War: Force and Political Responsibility*. New York: Charles Scribner's Sons, 1968.

Ratzinger, Cardinal. Congregation for the Doctrine of the Faith. *Instruction on Respect for Human Life in Its Origin and on the Dignity of Procreation: Replies to Certain Questions of the Day*. Rome: The Vatican, 1987.

Rawls, John. *A Theory of Justice*. Cambridge, MA: Harvard University Press, 1971.

———. *Political Liberalism*. New York: Columbia University Press, 1993.

Riener, Jacob and N. Stampfer. *Ethical Wills: A Modern Jewish Treasury*. New York: Schocken Books, 1983.

Rifkin, Jeremy. *The Biotech Century*. London: Gollancz, 1998.

Sagan, Carl. *The Demon-haunted World: Science as a Candle in the Dark*. New York: Random House, 1996.

Saunders, Cicely. "Foreword." *The Oxford Textbook of Palliative Medicine.* Oxford: Oxford University Press, 1993.

Schaeffer, Francis A. *The God Who Is There.* Downer's Grove, IL: InterVarsity Press, 1968.

Schweitzer, Albert. *On the Edge of the Forest.* Translated by C. T. Campion. New York: Macmillan, 1948.

Shattuck, R. *Forbidden Knowledge: From Prometheus to Pornography.* New York: St. Martin's Press, 1996.

Shelp, Earl E., ed. *Sexuality and Medicine Vol. II: Ethical Viewpoints in Transition.* Dordrecht, Holland: D. Reidel Publishing Co., 1987.

Simmons, Paul D. *Birth and Death: Bioethical Decision-making.* Philadelphia: Westminster Press, 1983.

———. "Theological Approaches to Sexuality: An Overview." In *Sexuality and Medicine Vol. II: Ethical Viewpoints in Transition.* Ed. Earl E. Shelp. Dordrecht, Holland: D. Reidel Publishing Co., 1987.

———. *Freedom of Conscience.* Amherst, NY: Prometheus, 2000.

———. *The Southern Baptists: Religious Beliefs and Healthcare Decisions.* Chicago: Park Ridge Center, 2002.

Smith, W. J. *Culture of Death: the Assault on Medical Ethics in America.* San Francisco: Encounter Books, 2000.

Soelle, D. *Suffering.* Trans. E. Kalin. Philadelphia: Fortress Press, 1975.

Stassen, G. H. "Critical Variables in Christian Social Ethics." In *Issues in Christian Ethics.* Ed. Paul D. Simmons. Nashville: Broadman Press, 1980.

Tillich, Paul. *The Eternal Now.* New York: Charles Scribner's Sons, 1963.

———. *Dynamics of Faith.* New York: Harper & Brothers, 1957.

Turnage, Mac and Anne. *Graceful Aging: Biblical Perspectives.* Atlanta: Presbyterian Office on Aging, 1984.

Veatch, R. M. "Justice, the Basic Social Contract, and Health Care." In *A Theory of Medical Ethics.* New York: Basic Books, 1981.

Verhey, Allan. *Reading the Bible in the Strange World of Medicine.* Grand Rapids, MI: Eerdmans Publishing Co., 2003.

Wachter, E. M. and K. G. Shojania. *Internal Bleeding: The Truth Behind America's Terrifying Epidemic of Medical Mistakes.* New York: Rugged Land, 2004.

Webster's New World Dictionary, 3rd ed. New York: Prentice Hall, 1991.

Wiesel, Elie. *The Oath.* New York: Random House, 1973.

Woodman, Marion. *The Pregnant Virgin: A Process of Psychological Transformation.* Toronto: Inner City Books, 1985.

Articles in Journals, Conference Presentations

Ai, A. L., R. E. Dunkle, C. Peterson, and S. F. Bolling. "The Role of Private Prayer in Psychological Recovery among Midlife and Aged Patients following Cardiac Surgery." *The Gerontologist* 38/5 (1998).

Ai, A. L., C. Peterson, S. F. Bolling, and H. Koenig, "Private Prayer and Optimism in Middle-aged and Older Patients Awaiting Cardiac Surgery." *The Gerontologist* 42/1 (2002).

Annas, G. J. "Prisoner in the I.C.U.: The Tragedy of William Bartling." *Hastings Center Report* 14/6 (December 1984): 28–29.

———. "Why We Should Ban Human Cloning." *NEJM* 339/2 (July 9, 1998).

———. "Reframing the Debate on Healthcare Reform by Replacing Our Metaphors." *NEJM* 332/11 (March 16, 1995).

Basgoz, N. and J. K. Preiksaitis. "Post-transplant lymphoproliferative disorder." *Infectious Disease Clinics of North America* (1995).

Baye, Bettye. "Ministers differ on whether churches should target abortion or poverty." *Courier-Journal* (Louisville KY), November 10, 2005. Editorial.

Berger, Philip A., Jack D. Barchas, and Thomas A. Gonda. "Schizophrenia: The Problem and Its Treatment." *The Stanford Magazine* (Fall/Winter 1980).

Bower, B. "Mind-expanding Machines." *Science News* 164 (August 30, 2003).

Breidenbach, W. C. et al. "The Ethics of the First Human Hand Transplantation." Louisville KY, Jewish Hospital, 1998. Unpublished

———. "Medicine at the Edges: Ethics and Innovation." Grand Rounds, Conference Center, Rudd Building, Jewish Hospital, Louisville, KY, August 20, 1999. Unpublished.

Brock, Dan W. "Voluntary Active Euthanasia." *Hastings Center Report* 22/2 (March/April 1992).

———. "How Much Is One Life Worth?" *Hastings Center Report* 36/3 (May/June 2006).

Callahan, Daniel. "Limiting Health Care for the Old?" *The Nation* (August 15, 1987).

———. "Rationing medical progress: the way to affordable health care." *NEJM* 322/25 (June 21, 1990).

Cassell, E. J. "The Nature of Suffering and the Goals of Medicine." *NEJM* 306/11 (1992).

"Computation Takes a Quantum Leap." *Science News* 158 (August 26, 2000).

Cooley, D. A. "Mechanical Circulatory Support Systems: Past, Present, and Future." *Annals of Thoracic Surgery* 68 (1999).

Cooney, William P. and V. R. Hentz, "Hand Transplantation—Primum Non Nocere." Position Statement of the Council of the American Society for Surgery of the Hand. *The Journal of Hand Surgery* 27A/1 (January 2002).

Copeland, J. G. et al. "The CardioWest Total Artificial Heart Bridge to Transplantation: 1993–1996 National Trial." *Annals of Thoracic Surgery* 66 (1998).

———. "Current Status and Future Directions for a Total Artificial Heart with a Past." *Artificial Organs* 22/11 (1998).

Cranford, Ron and Lawrence Gostin. "Futility: A Concept in Search of a Definition." *Law, Medicine & Health Care* 20/4 (Winter 1992).

Cruzan v. Director, Missouri Department of Health, 110 S. Ct. 2841 (1990).

Declaration of Helsinki. World Medical Association, 48th General Assembly, West Somerset, Republic of South Africa, 1996.

Dyck, Arthur. "Physician-assisted Suicide—Is It Ethical?" *Harvard Divinity Bulletin* 21/4 (1992).

Eisenberg, Leon. "The social imperatives of medical research" *Science* 198 (December 16, 1977): 1105–10.

Eisner, Thomas. "Chemical Ecology and Genetic Engineering: The Prospects for Plant Protection and the Need for Plant Habitat Conservation." Symposium on Tropical Biology and Agriculture, St. Louis, MO, Monsanto Company, July 15, 1986.

Fleck, Leonard. "The Costs of Caring: Who Pays? Who Profits? Who Panders? *Hastings Center Report* 36/3 (May/June 2006).

Gearhart, J. "New potential for human embryonic stem cells." *Science* 282 (1998): 1061–62.

Gruber, S. A. and J. A. Matas. "Etiology and pathogenesis of tumors occurring after organ transplantation." *Transplantation Science* 4 (1994): 87–104.

Hall, Elizabeth. "A Conversation with Clifford Grobstein." *Psychology Today* (September 1989).

Heidler, Timothy. "Patient perspectives." Second International Symposium on Composite Tissue Allotransplantation. Jewish Hospital, Louisville, KY, May 18, 2000.

Hofmann, G. O. "Knee and femur transplantation." Second International Symposium on Composite Tissue Allotransplantation. Jewish Hospital, Louisville, KY, May 18, 2000.

Honda, N. et al. "Ultracompact, Completely Implantable Permanent Use Electromechanical Ventricular Assist Device and Total Artificial Heart." *Artificial Organs* 23/3 (1999).

Jones, Neil. "Histological, immunological and functional comparison of nerve allografts immunosuppressed with cyclosporine, FK–506, RS–61443 and rapamycin." Second International Symposium on Composite Tissue Allotransplantation. Jewish Hospital, Louisville, KY, May 19, 2000.

Joy, Bill. "Why the Future Does Not Need Us," 1 at www.wired.com/wired/archive/8.04/joy.html.

Juengst, E. and M. Fossel. "The Ethics of Embryonic Stem Cells—Now and Forever, Cells without End." *JAMA* 284/24 (December 27, 2000).

Kass, Leon R. "Is There a Right to Die?" *Hastings Center Report* 23 (January/February 1993).

Kassirer, J. P. "Managed Care and the Morality of the Marketplace." *NEJM* 333/1 (July 6, 1995): 50–52.

Lamm, Richard D. "Health Care as Economic Cancer." *Dialysis and Transplantation* 16 (1987).

Leshner, Alan I. "Don't Let Ideology Trump Science." *Science* 302/5650 (November 28, 2003).

Luo, Michael. "On Abortion: It's the Bible of Ambiguity." *The New York Times* Week in Review (November 13, 2005).

Lynn, Joanne and James Childress, "Must Patients Always Be Given Food and water?" *Hastings Center Report* 13/5 (October 1983).

Malinowski, M. J. "Capitation, Advances in Medical Technology, and the Advent of a New Era in Medical Ethics." *American Journal of Law & Medicine* 22/2 & 3 (1996): 339.

Marquis, Don. "Abortion and the Beginning and End of Human Life." *Journal of Law, Medicine & Ethics* 34/1 (Spring 2006).

May, W. F. "Who Cares for the Elderly?" *Hastings Center Report* 12/6 (December 1982).

McKenzie, David. "Church, State and Physician-assisted Suicide." *Journal of Church and State* 46/4 (Autumn 2004).

Messikomer, C. M., R. C. Fox, and J. P. Swazey. "The Presence and Influence of Religion in American Bioethics." *Perspectives in Biology and Medicine* 44/4 (Autumn 2001).

Midgley, Mary. "Biotechnology and Monstrosity." *Hastings Center Report* 30/5 (September/October 2000).

Moore, Francis D. “Prolonging Life, Permitting Life to End.” *Harvard Magazine* 97/6 (July/August 1995).

_______. “Three ethical revolutions: ancient assumptions remodeled under pressure of transplantation.” *Transplantation Proceedings* 20 (1988): 1061–67.

Murray, Thomas. “Gifts of the Body and the Needs of Strangers.” *Hastings Center Report* (April 1987); and in Carole Levine, ed. *Taking Sides: Clashing Views on Controversial Bioethical Issues*, 7th ed. Guilford, CT: Dushkin, 1997, 350–58.

Pedersen, R. A. “Embryonic stem cells for medicine.” *Scientific American* (1999).

Pellegrino, Edmund P. “The Metamorphosis of Medical Ethics: A 30-year Perspective.” *JAMA* 269/9 (March 3, 1993).

Press, Eyal. “My Father’s Abortion War.” *The New York Times Magazine* (January 22, 2006).

Quill, Timothy. “Death and Dignity: A Case of Individualized Decision-making.” *NEJM* 324/10 (March 7, 1991).

———. “Doctor, I want to die. Will you help me?” *JAMA* (August 18, 1993).

Quill, Timothy and Diane Meier. “The Big Chill—Inserting the DEA into End-of-life Care.” *NEJM* 354/1 (January 5, 2006).

Ramsey, Paul. “Shall we ‘Reproduce’? I. The medical ethics of in vitro fertilization.” *JAMA* 220/10 (June 5, 1972): 1347.

Ruane, Mebd. “The Irish Referendum: The End of Rome Rule.” *Conscience* 23/1 (Spring 2002).

Rushton, Cindy H. and Kathleen Russell. “The Language of Miracles: Ethical Challenges.” *Pediatric Nursing* (January/February, 1996).

Schiff, G. D. et al. “A Better-Quality Alternative Single-Payer National Health System Reform.” *JAMA* 272/10 (September 14, 1994).

Schneiderman, Lawrence, Nancy Jecker, and Albert Jonsen. “Medical Futility: Its Meaning and Ethical Implications.” *Annals of Internal Medicine* 112 (1990): 949–54.

Seigler, Mark. “Ethical issues in innovative surgery: should we attempt a cadaveric hand transplantation in a human subject?” *Transplantation Proceedings* 30 (1998): 2779–82.

Sharlet, Jeff. “Through a Glass, Darkly: How the Christian Right Is Reimagining U.S. History.” *Harper’s Magazine* (December 2006).

Simmons, Paul D. “Post-abortion Depression & the Ethics of Truth-telling.” *Christian Ethics Today* 9/3 (Summer 2003).

———. "Religious Liberty and the Abortion Debate." *Journal of Church and State* 32 (Summer 1990).

———. "Religious Liberty and Abortion Policy: *Casey* as 'Catch–22.'" *Journal of Church and State* 42 (Winter 2000).

———. "Ethical considerations of artificial heart implantations." *Annals of Clinical and Laboratory Science* (1986).

Sinsheimer, Robert. "The Prospect of Designed Genetic Change." *Engineering and Science* (April 1969).

Soros, George. "The Capitalist Threat." *The Atlantic Monthly* (February 1997).

"Stem Cell Surprise." *Science News* 163/9 (March 1, 2003): 131.

Steinbock, Bonnie. "The Morality of Killing Human Embryos." *The Journal of Law, Medicine & Ethics* 34/1 (Spring 2006): 26–34.

Strome, M. "Larynx transplantation." Second International Symposium on Composite Tissue Allotransplantation, Jewish Hospital, Louisville, KY, May 18, 2000.

Stubblefield, P. G. and D. A. Grimes. "Septic Abortion," *NEJM* 331 (1994): 310–14.

Swanson, Jeffrey and S. Van McCrary. "Doing All They Can: Physicians Who Deny Medical Futility." *Journal of Law, Medicine & Ethics* 22/4 (Winter 1994).

Tasca, R. J. and M. E. McClure. "The Emerging Technology and Application of Pre- implantation Genetic Diagnosis." *Journal of Law, Medicine and Ethics* 26 (1990).

Van Hooft, S. "The Meanings of Suffering." *Hastings Center Report* 28/5 (September/October 1998).

Vandekieft, G. K. "Breaking Bad News." *American Family Physician* (December 15, 2001).

Veatch, Robert M. "Abandoning informed consent." *Hastings Center Report* 25/2 (March–April 1995).

Verhey, Allen. "Choosing Death: The Ethics of Assisted Suicide." *The Christian Century* (July 17–24, 1996): 716–19.

Wanzer, S. et al. "The Physician's Responsibility Toward the Hopelessly Ill." *NEJM* 320/13 (March 30, 1989): 844–49.

Wiggins, Osbourne P. et al. "On the Ethics of Facial Transplantation Research." *American Journal of Bioethics* 4/3 (2004).

INDEX